Student Solutions Manual

to accompany

College Mathematics

2009 Update with MyMathLab®

Cheryl Cleaves
Southwest Tennessee Community College

Margie Hobbs
The University of Mississippi

PEARSON

Prentice
Hall

Upper Saddle River, New Jersey
Columbus, Ohio

Editor-in-Chief: Vernon Anthony
Senior Acquisitions Editor: Gary Bauer
Editorial Assistant: Megan Heintz
Project Manager: Rex Davidson
Senior Operations Supervisor: Pat Tonneman
Art Director: Diane Ernsberger
Cover Designer: Candace Rowley
Director of Marketing: David Gesell
Marketing Manager: Leigh Ann Sims

This book was set by TAB Publishing Services, Sunbury, Ohio. It was printed and bound by Demand Production Center. The cover was printed by Demand Production Center.

Pearson Prentice Hall[TM] is a trademark of Pearson Education, Inc.
Pearson® is a registered trademark of Pearson plc
Prentice Hall® is a registered trademark of Pearson Education, Inc.

Pearson Education Ltd., London
Pearson Education Singapore Pte. Ltd.
Pearson Education Canada, Ltd.
Pearson Education–Japan

Pearson Education Australia Pty. Limited
Pearson Education North Asia Ltd., Hong Kong
Pearson Educación de Mexico, S.A. de C.V.
Pearson Education Malaysia Pte. Ltd.

10 9 8 7 6 5 4 3 2 1

ISBN-13: 978-0-13-502523-9
ISBN-10:   0-13-502523-0

Contents

TO THE STUDENT

This manual contains the step-by-step solutions for odd-numbered exercises for the Chapter Review and Practice Test for each chapter. It also includes the solutions for odd-numbered exercises in each Cumulative Practice Test. The purpose of this manual is to assist you with your study of *College Mathematics: 2009 Update with MyMathLab*®. It is advisable to make several attempts to work the exercises before referring to the solutions given here. You will also want to study the examples and explanations given in the text in your process of understanding the concepts presented in the exercises.

A given exercise often can be solved many different ways. Some approaches may require more steps than others. The approach presented in this manual will not necessarily be the shortest way of working the exercise or the way you may have attempted to work the exercise. If your approach is not exactly like the approach presented in this manual, compare your approach with the approach presented here to determine if both processes employ sound reasoning strategies. The authors have made every attempt to verify the accuracy of the solutions. However, if you should determine that a solution in this manual is incorrect or incomplete, the authors would appreciate hearing your comments, which may be sent to the publishing address listed in this manual.

Mathematics skills are normally developed in a particular sequence so that new skills are reinforced and they build on previously learned skills; therefore, it is *very* important to master each skill before proceeding to the next skill.

Developing Your Study Plan

To be successful in a mathematics program, you must *practice* all required skills. The amount of practice needed varies from student to student, but the text is designed with built-in checks to determine if you have practiced sufficiently to master each skill.

Many books are available that offer suggestions for effective strategies for studying mathematics. In addition, you may work with your instructor to develop your study plan. Your study plan should utilize the resources that are available to you such as learning centers, tutors, computer programs, and videotapes.

You will discover that working with a study partner or participating in a study group is a very successful strategy for studying mathematics. An effective study plan can prove to be as important to your success in mathematics as your mastery of specific mathematics skills.

Using Your Calculator

The text gives many opportunities for you to develop your skills in using both the scientific and graphing calculators. A calculator is most effective when you estimate or predict the answer before you make the calculation. Developing your number sense about calculations will enable you to determine when you have made errors in sequencing operations or pressing the appropriate keys. A calculator should not replace your computations of single digit number facts and other easy calculations that can be performed mentally. The calculator, when used properly, can be an enormous tool in discovering and verifying mathematical concepts. Your proficiency in using the calculator will be desirable in most employment settings.

We hope you are successful in your study of mathematics!

Cheryl Cleaves
Margie Hobbs

Review of Basic Concepts

Chapter Review Exercises

1. (a) $\dfrac{3}{10} = 0.3$

A fraction with a denominator that is a power of 10 can be written as a decimal number by writing the numerator and placing the decimal in the appropriate place to indicate the proper place value of the denominator.

(b) $\dfrac{15}{100} = 0.15$

Since the denominator is 100, the last digit of the numerator is in the hundredths place.

(c) $\dfrac{4}{100} = 0.04$

Since the denominator is 100, the last digit is in the hundredths place.

3. (a) nearest hundred ④ _68

500

(b) nearest ten thousand ④ _9,238

50,000

(c) nearest tenth 41.③ _78

41.4

3 is in the tenths place. 7 is digit to the right, so add 1 to the 3. Digits to the left remain the same, digits to the right (and to the right of the decimal point) are dropped.

(d) nearest hundredth 6.8 ⑨ _57

6.90

(e) nearest ten-thousandth 23.460 ⑨ _7

23.4610

5. larger 4.783   4.79

↑   ↑

Compare each place value, left to right, until two digits in the same place are different, and compare those digits.

4.783 < ④.79 ; 4.79 is larger.

7. smallest to largest

0.021   0.0216   0.02
  ↑   ↑   ↑
0.021   0.0216   0.020
  ↑   ↑   ↑

0.020 is smaller, now compare 0.021   0.0216
                                 ↑   ↑
                          0.0210 < 0.0216
                              ↑   ↑

0.020 < 0.0210 < 0.0216
0.02 < 0.021 < 0.0216

9. $\dfrac{7}{8} = 0.875$ $\dfrac{6}{7} = 0.857$
              ↑   ↑

0.875 > 0.857

$\dfrac{7}{8}$ is larger than $\dfrac{6}{7}$.

11. (a) $8 + 5 + 3 + 6 + 2 + 4 = 28$
         (b) $7 + 4 + 3 + 2 + 5 + 4 = 25$

13.
$$\begin{array}{r} {}^{1\,1} \\ 10.4 \\ 15.3 \\ 2.9 \\ + \ 6.3 \\ \hline 34.9 \end{array}$$
34.9 is the total number of kilowatts used.

15. (a) $\begin{array}{r} 21.34 \\ -\,16.73 \\ \hline 4.61 \end{array}$ (b) $\begin{array}{r} 15.934 \\ -\,12.807 \\ \hline 3.127 \end{array}$

(c) $\begin{array}{r} 284.730 \\ -\ \ 79.831 \\ \hline 204.899 \end{array}$ (d) $\begin{array}{r} 13,342 \\ -\ \ 1,202 \\ \hline 12,140 \end{array}$

17. $\begin{array}{r} 8.296 \\ -\,0.005 \\ \hline 8.291 \text{ in.} \end{array}$ $\begin{array}{r} 8.296 \\ +\,0.005 \\ \hline 8.301 \text{ in.} \end{array}$

The limit dimensions are 8.291 in. to 8.301 in.

19. B $\begin{array}{r} {}^{1\ \ 1} \\ 1.861 \\ C\ +\,1.946 \\ \hline 3.807 \end{array}$ D $\begin{array}{r} 4.237 \\ -\,3.807 \\ \hline A\ \ \ 0.430 \end{array}$

The length of A is 0.43 in.

21. $\begin{array}{r} 8.935 \\ -\,0.005 \\ \hline 8.930 \text{ in.} \end{array}$ $\begin{array}{r} {}^{1} \\ 8.935 \\ +\,0.005 \\ \hline 8.940 \text{ in.} \end{array}$

The limit dimensions of D are 8.930 in. to 8.940 in.

23. $2(6)(7) = 12(7) = 84$

$$\begin{array}{r} 12 \\ \times\ \ 7 \\ \hline 84 \end{array}$$

25.
$$\begin{array}{r} 305 \\ \times\ 45 \\ \hline 1525 \\ 1220\ \ \\ \hline 13,725 \end{array}$$

27.
$$\begin{array}{r} {}^{4}_{2}\ \ {}^{1} \\ 56,002 \\ \times\ \ \ \ 704\ 0 \\ \hline 224008 \\ 3920140\ \ \ \\ \hline 394,254,080 \end{array}$$

29.
$$\begin{array}{r} \$\ \ 67 \\ \times\ 21 \\ \hline 67 \\ 134\ \ \\ \hline \$\,1,407 \end{array}$$
A business would pay \$1,407 for the keyboards.

31.

$$\begin{array}{r} \$\;\;\;305 \\ \times\;144 \\ \hline 1220 \\ 1220\;\; \\ 305\;\;\; \\ \hline \$43{,}920 \end{array}$$

The dealer paid \$43,920 for the order.

33. $\$365 \times 36$

Estimate	Exact	Check
$\$370 \times 40 = \$14{,}800$		

$$\begin{array}{r} 365 \\ \times\;\;\;\;36 \\ \hline 2190 \\ 1095\;\; \\ \hline \$13{,}140 \end{array} \qquad \begin{array}{r} 36 \\ \times\;\;365 \\ \hline 180 \\ 216\;\; \\ 108\;\;\; \\ \hline 13{,}140 \end{array}$$

The worker will earn \$13,140.

35.

Estimate $\qquad$ Exact

$A = l \times w \qquad A = l \times w$
$= 1{,}900 \times 600 \qquad = 1{,}940.7 \times 620.4$
$= 1{,}140{,}000 \text{ ft}^2 \qquad = 1{,}204{,}010.28 \text{ ft}^2$

The area of the land is 1,204,010.28 square feet.

37.

$$\begin{array}{r} 0.00014 \\ \times\;\;\;\;\;864 \\ \hline 56 \\ 84\;\; \\ 112\;\;\; \\ \hline 0.12096 \end{array}$$

The tape will expand 0.12096 in.

39. $325 \div 25$

$$\begin{array}{r} 13 \\ 25\overline{)325} \\ 25\;\; \\ \hline 75 \\ 75 \\ \hline 0 \end{array}$$

41. $30{,}126 \div 15$

$$\begin{array}{r} 2{,}008.4 \\ 15\overline{)30{,}126.0} \\ 30\;\;\;\;\;\;\;\;\; \\ \hline 1\;\;\;\;\;\;\; \\ 0\;\;\;\;\;\;\; \\ \hline 12\;\;\;\;\; \\ 0\;\;\;\;\; \\ \hline 126\;\;\; \\ 120\;\;\; \\ \hline 60\; \\ 60\; \\ \hline 0 \end{array}$$

43.

$$\begin{array}{r} 23 \text{ R } 11 \\ 27\overline{)632} \\ 54\;\; \\ \hline 92 \\ 81 \\ \hline 11 \end{array}$$

Each volunteer will receive 23 envelopes and 11 will be left over.

45. average, nearest hundredth

$$\begin{array}{r} {}^{1\,3\;\,2} \\ 42.34 \\ 38.97 \\ 51.95 \\ +\;61.88 \\ \hline 195.14 \end{array} \qquad \begin{array}{r} 48.785 \approx 48.79 \text{ ft} \\ 4\overline{)195.140} \\ 16\;\;\;\;\;\;\;\; \\ \hline 35\;\;\;\;\;\; \\ 32\;\;\;\;\;\; \\ \hline 31\;\;\;\;\; \\ 28\;\;\;\;\; \\ \hline 34\;\;\; \\ 32\;\;\; \\ \hline 20\; \\ 20\; \\ \hline 0 \end{array}$$

47. (a)   base $\rightarrow 7^{3 \leftarrow \text{exponent}}$

         $7^3 = 7 \times 7 \times 7 = 343$

     (b)   base $\rightarrow 2.3^{4 \leftarrow \text{exponent}}$

         $2.3^4 = 2.3 \times 2.3 \times 2.3 \times 2.3 = 27.9841$

     (c)   base $\rightarrow 8^{4 \leftarrow \text{exponent}}$

         $8^4 = 8 \times 8 \times 8 \times 8 = 4{,}096$

49. (a) $1^2 = 1 \times 1 = 1$

     (b) $125^2 = 125 \times 125 = 15{,}625$

     (c) $5.6^2 = 5.6 \times 5.6 = 31.36$

     (d) $21^2 = 21 \times 21 = 441$

     some calculators: 21 $\boxed{x^2}$

     other calculators: 21 $\boxed{x^2}$ $\boxed{=}$

51. $10 = 10^1$
$1{,}000 = 10^3$
$10{,}000 = 10^4$
$100{,}000 = 10^5$

53. (a) $700 \div 100 = 7$

     (b) $40.56 \div 1{,}000 = 0.04056$

     (c) $60.5 \div 100 = 0.605$

     (d) $23{,}079 \div 10{,}000 = 2.3079$

55.

$4^2 \cdot (12 - 7) - 8 + 3 =$	Do operation within parentheses first.
$4^2 \cdot 5 - 8 + 3 =$	Evaluate exponentiation.
$16 \cdot 5 - 8 + 3 =$	Multiply.
$80 - 8 + 3 =$	Add and subtract from left to right.
$72 + 3 =$	
75	

57.

$5 + 21 \div 3 \cdot 7 =$	Divide.
$5 + 7 \cdot 7 =$	Multiply.
$5 + 49 =$	Add.
54	

59.

$21 + 7 \cdot 2 - 5 \cdot 4 =$	Multiply left to right.
$21 + 14 - 20 =$	Add and subtract from left to right.
$35 - 20 =$	Subtract.
15	

61.

$18 - 5 \cdot 2 + 7 =$	Multiply.
$18 - 10 + 7 =$	Add and subtract from left to right.
$8 + 7 =$	Add.
15	

63.

$5 - 2 \cdot 2 + 12 =$	Multiply.
$5 - 4 + 12 =$	Add and subtract from left to right.
$1 + 12 =$	Add.
13	

65.

$3.1 \cdot 4 \cdot \sqrt{16} - 6^2 =$	Evaluate exponentiation and square root from left to right.
$3.1 \cdot 4 \cdot 4 - 6^2 =$	Evaluate exponentiation.
$3.1 \cdot 4 \cdot 4 - 36 =$	Multiply from left to right.
$12.4 \cdot 4 - 36 =$	Multiply.
$49.6 - 36 =$	Subtract.
13.6	

67. $5.2^3 - \sqrt{81} \cdot (2 + 1) =$ Do operations within parentheses first.

       $5.2^3 - \sqrt{81} \cdot 3 =$ Evaluate exponentiation and square roots from left to right.

     $140.608 - \sqrt{81} \cdot 3 =$ Evaluate square root.

      $140.608 - 9 \cdot 3 =$ Multiply.

       $140.608 - 27 =$ Subtract.

         113.608

69. $4 + 5 - 2 \cdot 3 =$ Multiply.

     $4 + 5 - 6 =$ Add and subtract from left to right.

      $9 - 6 =$ Subtract.

        3

71. $12 \div 4(6) =$ Divide first.

      $3(6) =$ Multiply.

       18

73. $27 \div 3(14 - 8) - 2 + 3^2 =$ Do operations inside parentheses first.

     $27 \div 3(6) - 2 + 3^2 =$ Evaluate exponentiation.

     $27 \div 3(6) - 2 + 9 =$ Multiply and divide from left to right.

       $9(6) - 2 + 9 =$ Multiply.

         $54 - 2 + 9 =$ Add and subtract from left to right.

           $52 + 9 =$ Add.

            61

75. $584 \div 12 = 48.67$

            or 49 boxes

77. $48 \cdot 100 \cdot \$1.40 = \$6,720$

     $40 \cdot 110 \cdot \$1.40 = \$6,160$

      Total Rent $= \$12,880$

79. $(43 + 68 + 72 + 59 + 21) \div 5 = \$263 \div 5 = \$52.60$

      $\$52.60$ is the average cost per pair of shoes.

81.

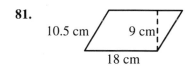

Perimeter $P = 2(b + s)$

          $= 2(18 + 10.5)$

          $= 2(28.5)$

          $= 57$ cm

83.   35 mm, 70 mm

$P = 2(l + w)$

   $= 2(70 + 35)$

   $= 2(105)$

   $= 210$ mm

85.   7.2 m, 7.2 m

$P = 4s$

   $= 4(7.2) = 28.8$ m

87.   15 ft, 18 ft

$P = 2(b) + 2(s)$

   $= 2(18) + 2(15)$

   $= 36 + 30 = 66$ ft

89.

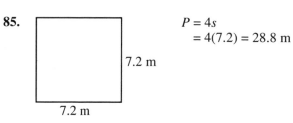

$P = 2(l) + 2(w)$

   $= 2(84) + 2(60)$

   $= 168 + 120 = 288$ in.

91.

$A = bh$
$A = 18(9)$
$A = 162 \text{ cm}^2$

$h = 9$ cm
$b = 18$ cm

93.

$A = lw$
$A = 70(35)$
$A = 2{,}450 \text{ mm}^2$

35 mm
70 mm

95.

$A = s^2$
$A = (7.2)^2$
$A = 51.84 \text{ m}^2$

7.2 m
7.2 m

97. $A = lw$
$A = 275(120)$
$A = 33{,}000 \text{ ft}^2$

99. $A = lw$
$A = 18(16.5)$
$A = 297 \text{ ft}^2$
$A = \dfrac{297 \text{ ft}^2}{1}\left(\dfrac{1 \text{ yd}^2}{9 \text{ ft}^2}\right) = 33 \text{ yd}^2$

101. $A = lw$
$A = 20(16)$
$A = 320 \text{ ft}^2$
$3(\$4.75) = \$14.25$

$$\begin{array}{r} 2.1 \\ 150\overline{)320.0} \\ \underline{300} \\ 20\ 0 \end{array} \; ; \; 3 \text{ gal needed}$$

Chapter 1 Practice Test

1. smaller: 5.09   5.1
           ↑   ↑
       ⎛5.09⎞ < 5.1

3. nearest tenth   48.③284
                        48.3

5.
$$\begin{array}{r} ^{2\,2} \\ 37 \\ 158 \\ 764 \\ +\ \ 48 \\ \hline 1{,}007 \end{array}$$

7.
$$\begin{array}{r} ^{2\,1}\ \ ^{4} \\ ^{1} \\ 13{,}207 \\ \times\ \ \ \ 702 \\ \hline ^{1\ 1} \\ 26414 \\ 924490 \\ \hline \$9{,}271{,}314 \end{array}$$

9. $3^2 + 5^3 =$　Evaluate exponentiation from left to right.
        $9 + 5^3 =$　Evaluate exponentiation.
        $9 + 125 =$　Add.
        134

11. $3 \times 6^2 - 4 \div 2 =$　Evaluate exponentiation.
        $3 \times 36 - 4 \div 2 =$　Multiply and divide from left to right.
        $108 - 4 \div 2 =$　Divide.
        $108 - 2 =$　Subtract.
        106

13.

Estimate:	Exact:
$480	$475
− 170	− 165
$310	$310

There is $310 left from his paycheck.

15. Estimate: Exact:

$$\begin{array}{r} \$10 \\ 10\overline{)\$100} \end{array}$$

$$\begin{array}{r} \$14 \\ 9\overline{)\$126} \\ \underline{9} \\ 36 \\ \underline{36} \\ 0 \end{array}$$

The cost per pizza is $14.

17. $42.73 \times 1,000 = 42.730 = 42,730$

19.

$$\begin{array}{r} 11.58 \approx 11.6 \text{ (nearest tenth)} \\ 7.2\overline{)83.410} \\ \underline{72} \\ 114 \\ \underline{72} \\ 421 \\ \underline{360} \\ 610 \\ \underline{576} \\ 34 \end{array}$$

21.

$$\begin{array}{r} 2 \\ 82 \\ 95 \\ 76 \\ 84 \\ 72 \\ + \ 91 \\ \hline 500 \end{array}$$

$$\begin{array}{r} 83.3 \approx 83 \text{ (nearest whole number)} \\ 6\overline{)500.0} \\ \underline{48} \\ 20 \\ \underline{18} \\ 20 \\ \underline{18} \\ 2 \end{array}$$

23. Estimate, nearest tenth.

$$\begin{array}{r} 0.87 \\ - \ 0.328 \end{array}$$

$$\begin{array}{r} 0.9 \\ - \ 0.3 \\ \hline 0.6 \end{array}$$

25. $\$1.75 \times 10,000 = 1.7500 = \$17,500$

The heating oil costs $17,500.

27. $P = 2(l + w)$
$P = 2(24.5 + 21)$
$P = 2(45.5)$
$P = 91$ ft
$A = lw$
$A = 24.5(21)$
$A = 514.5$ ft^2

29. $P = 2(b + s)$
$P = 2(3.8 + 2.3)$
$P = 2(6.1)$
$P = 12.2$ in.
$A = bh$
$A = 3.8(1.6)$
$A = 6.08$ in^2

chapter 2 Review of Fractions

Chapter Review Exercises

1. $24 = 12(2)$; $36 = 12(3)$; $48 = 12(4)$; $60 = 12(5)$; $72 = 12(6)$; answers may vary.

3. $20 = 10(2)$; $30 = 10(3)$; $40 = 10(4)$; $50 = 10(5)$; $60 = 10(6)$; answers may vary.

5. $42 = 21(2)$; $63 = 21(3)$; $84 = 21(4)$; $105 = 21(5)$; $126 = 21(6)$; answers may vary.

7. $14 = 7(2)$; $21 = 7(3)$; $28 = 7(4)$; $35 = 7(5)$; $42 = 7(6)$; answers may vary.

9. $16 = 8(2)$; $24 = 8(3)$; $32 = 8(4)$; $40 = 8(5)$; $48 = 8(6)$; answers may vary.

11. Yes; sum of digits is divisible by 3.　　**13.** No; divisible by 2, but not by 3.

15. Yes; last digit is 0.　　**17.** No; sum of digits is not divisible by 9.

19. 1, 48　1, 2, 3, 4, 6, 8, 12, 16, 24, 48
　　2, 24
　　3, 16
　　4, 12
　　6, 8

21. 1, 51　1, 3, 17, 51
　　3, 17

23. 1, 74　1, 2, 37, 74
　　2, 37

25. 18, composite　$1 \cdot 18$
　　　　　　　　$2 \cdot 9$
　　　　　　　　$3 \cdot 6$

27. 21, composite　$1 \cdot 21$
　　　　　　　　$3 \cdot 7$

29. $2 \cdot 3 \cdot 7$

31. $2 \cdot 7 \cdot 7$ or $2 \cdot 7^2$

33. $18 = 2 \cdot 3^2$
　　$40 = 2^3 \cdot 5$
　　$\text{LCM} = 2^3 \cdot 3^2 \cdot 5$
　　　　$= 360$

35. $12 = 2^2 \cdot 3$
　　$18 = 2 \cdot 3^2$
　　$30 = 2 \cdot 3 \cdot 5$
　　$\text{LCM} = 2^2 \cdot 3^2 \cdot 5$
　　　　$= 180$

37. $10 = 2 \cdot 5$
$12 = 2^2 \cdot 3$
$GCF = 2$

39. $12 = 2^2 \cdot 3$
$18 = 2 \cdot 3^2$
$30 = 2 \cdot 3 \cdot 5$
$GCF = 2 \cdot 3$
$= 6$

41. $\dfrac{5}{8} = \dfrac{?}{24}$

$8\overline{)24}^{\,3}, \; 3 \times 5 = 15$

$\dfrac{5}{8} = \dfrac{15}{24}$

43. $\dfrac{5}{12} = \dfrac{?}{60}$

$12\overline{)60}^{\,5} \; 5 \times 5 = 25$

$\dfrac{5}{12} = \dfrac{25}{60}$

45. $\dfrac{2}{3} = \dfrac{?}{15}$

$3\overline{)15}^{\,5}, \; 5 \times 2 = 10$

$\dfrac{2}{3} = \dfrac{10}{15}$

47. $\dfrac{3}{4} = \dfrac{?}{32}$

$4\overline{)32}^{\,8}, \; 8 \times 3 = 24$

$\dfrac{3}{4} = \dfrac{24}{32}$

49. $\dfrac{1}{5} = \dfrac{?}{55}$

$5\overline{)55}^{\,11}, \; 11 \times 1 = 11$

$\dfrac{1}{5} = \dfrac{11}{55}$

51. $\dfrac{4}{5} = \dfrac{?}{20}$

$5\overline{)20}^{\,4}, \; 4 \times 4 = 16$

$\dfrac{4}{5} = \dfrac{16}{20}$

53. $\dfrac{6}{12} = \dfrac{6 \div 6}{12 \div 6} = \dfrac{1}{2}$

55. $\dfrac{4}{32} = \dfrac{4 \div 4}{32 \div 4} = \dfrac{1}{8}$

57. $\dfrac{2}{8} = \dfrac{2 \div 2}{8 \div 2} = \dfrac{1}{4}$

59. $\dfrac{34}{64} = \dfrac{34 \div 2}{64 \div 2} = \dfrac{17}{32}$

61. $\dfrac{12}{32} = \dfrac{12 \div 4}{32 \div 4} = \dfrac{3}{8}$

63. $\dfrac{6}{8} = \dfrac{6 \div 2}{8 \div 2} = \dfrac{3}{4}$

65. The line segment is past 5 in. but before 6 in.; thus the measure will be a mixed number. $5\dfrac{1}{4}$ inch

67. $4\dfrac{7}{16}$ inch   **69.** $3\dfrac{15}{16}$ inch   **71.** $3\dfrac{9}{16}$ inch   **73.** $2\dfrac{3}{4}$ inch   **75.** $1\dfrac{1}{2}$ in.

77. $2\dfrac{1}{8}$ in.   **79.** $0.7 = \dfrac{7}{10}$ **81.** $0.95 = \dfrac{95 \div 5}{100 \div 5} = \dfrac{19}{20}$

83. $0.872 = \dfrac{872 \div 8}{1{,}000 \div 8} = \dfrac{109}{125}$ **85.** $0.02 = \dfrac{2 \div 2}{100 \div 2} = \dfrac{1}{50}$

87. $5\overline{)1.0}$ with 0.2 above, $\underline{1.0}$ $\dfrac{1}{5} = 0.2$

89. $8\overline{)5.000}$ with 0.625 above $\dfrac{5}{8} = 0.625$
$\underline{48}$
20
$\underline{16}$
40
$\underline{40}$

91. $11\overline{)9.0000}$ with 0.8181 above $\dfrac{9}{11} = 0.818$ (rounded)
$\underline{88}$
20
$\underline{11}$
90
$\underline{88}$
20
$\underline{11}$
9

93. $\dfrac{18}{5} = 5\overline{)18} = 3\dfrac{3}{5}$ with 3 above; $\underline{15}$; 3

95. $\dfrac{39}{8} = 8\overline{)39} = 4\dfrac{7}{8}$ with 4 above; $\underline{32}$; 7

97. $\dfrac{43}{8} = 8\overline{)43} = 5\dfrac{3}{8}$ with 5 above; $\underline{40}$; 3

99. $\dfrac{175}{2} = 2\overline{)175} = 87\dfrac{1}{2}$ with 87 above; $\underline{16}$; 15 ; $\underline{14}$; 1

101. $\dfrac{18}{12} = 12\overline{)18} = 1\dfrac{6}{12} = 1\dfrac{1}{2}$ with 1 above; $\underline{12}$; 6

103. $8 = \dfrac{8}{1}$

105. $7\dfrac{1}{8} = \dfrac{(8 \times 7) + 1}{8} = \dfrac{57}{8}$

107. $9\dfrac{3}{16} = \dfrac{(16 \times 9) + 3}{16} = \dfrac{147}{16}$

109. $4\dfrac{3}{5} = \dfrac{(5 \times 4) + 3}{5} = \dfrac{23}{5}$

111. $12 = \dfrac{12}{1}$

113. $5\dfrac{1}{3} = \dfrac{(3 \times 5) + 1}{3} = \dfrac{16}{3}$

115. $2 = \dfrac{?}{10}$

$2 = \dfrac{2}{1} = \dfrac{2 \times 10}{1 \times 10} = \dfrac{20}{10}$

117. $11 = \dfrac{?}{3}$

$11 = \dfrac{11}{1} = \dfrac{11 \times 3}{1 \times 3} = \dfrac{33}{3}$

119. $\dfrac{1}{4}, \dfrac{1}{3}, \dfrac{1}{5}$ Factor each denominator.
$2 \times 2 \quad 3 \quad 5$
2^2
$LCD = 2^2 \times 3 \times 5 = 4 \times 3 \times 5 = 60$

121. $\dfrac{3}{4}, \dfrac{1}{16}$ Factor each denominator.
$2 \times 2 \quad 2 \times 8$
$2^2 \quad 2 \times 2 \times 4$
$2 \times 2 \times 2 \times 2$
2^4
$LCD = 2^4 = 16$

123. $\dfrac{5}{12}, \dfrac{3}{10}, \dfrac{13}{15}$ Factor each denominator.
$2 \times 6 \quad 2 \times 5 \quad 3 \times 5$
$2 \times 2 \times 3$
$2^2 \times 3$
$LCD = 2^2 \times 3 \times 5 = 4 \times 3 \times 5 = 60$

125. $\dfrac{5}{8} \quad \dfrac{9}{16}$

$\dfrac{10}{16} > \dfrac{9}{16}$ $\dfrac{5}{8} = \dfrac{5 \times 2}{8 \times 2} = \dfrac{10}{16}$

$\dfrac{5}{8} > \dfrac{9}{16}$

$\dfrac{5}{8}$ is greater than $\dfrac{9}{16}$

127. $\dfrac{3}{8} \overset{?}{<} \dfrac{1}{2}$

$\dfrac{3}{8} < \dfrac{4}{8}$ $\qquad \dfrac{1}{2} = \dfrac{1 \times 4}{2 \times 4} = \dfrac{4}{8}$

$\dfrac{3}{8} < \dfrac{1}{2}$

No, $\dfrac{3}{8}$ in. thick plaster board is not thicker than a $\dfrac{1}{2}$ in. thick piece.

129. $\dfrac{19}{32} \overset{?}{<} \dfrac{7}{8}$

$\dfrac{19}{32} < \dfrac{28}{32}$ $\qquad \dfrac{7}{8} = \dfrac{7 \times 4}{8 \times 4} = \dfrac{28}{32}$

$\dfrac{19}{32} < \dfrac{7}{8}$

A $\dfrac{19}{32}$ in. wrench is smaller than a $\dfrac{7}{8}$ in. bolt head.

131. $\dfrac{5}{8}, \dfrac{3}{8}$

$\dfrac{3}{8} < \dfrac{5}{8}$ or $\dfrac{5}{8} > \dfrac{3}{8}$

$\dfrac{3}{8}$ is smaller.

133. $\dfrac{3}{8} < \dfrac{4}{8}$

$\dfrac{3}{8}$ is smaller.

135. $\dfrac{1}{4} \qquad \dfrac{3}{16}$

$\dfrac{4}{16} > \dfrac{3}{16}$ $\qquad \dfrac{1}{4} = \dfrac{1 \times 4}{4 \times 4} = \dfrac{4}{16}$

$\dfrac{1}{4} > \dfrac{3}{16}$

$\dfrac{3}{16}$ is smaller.

137. $\dfrac{7}{8} \qquad \dfrac{27}{32}$

$\dfrac{28}{32} > \dfrac{27}{32}$ $\qquad \dfrac{7}{8} = \dfrac{7 \times 4}{8 \times 4} = \dfrac{28}{32}$

$\dfrac{7}{8} > \dfrac{27}{32}$

$\dfrac{27}{32}$ is smaller.

139. $\dfrac{1}{2} \qquad \dfrac{9}{19}$

$\dfrac{19}{38} > \dfrac{18}{38}$ $\qquad \dfrac{1}{2} = \dfrac{1 \times 19}{2 \times 19} = \dfrac{19}{38}, \dfrac{9}{19} = \dfrac{9 \times 2}{19 \times 2} = \dfrac{18}{38}$

$\dfrac{1}{2} > \dfrac{9}{19}$

$\dfrac{9}{19}$ is smaller.

143. $0.272 = \dfrac{272}{1,000} = \dfrac{272 \times 11}{1,000 \times 11} = \dfrac{2,992}{11,000}$

$\dfrac{3}{11} = \dfrac{3 \times 1,000}{11 \times 1,000} = \dfrac{3,000}{11,000}$

$\dfrac{2,992}{11,000} < \dfrac{3,000}{11,000}$; so, $\dfrac{3}{11}$ is larger.

or $\dfrac{3}{11} \approx 0.2727$

$0.272 < 0.2727$

141. $0.26 = \dfrac{26}{100}$

$\dfrac{3}{4} = \dfrac{3 \times 25}{4 \times 25} = \dfrac{75}{100}$ $\qquad$ or $\dfrac{3}{4} = \dfrac{75}{100} = 0.75$

$0.75 > 0.26$

$\dfrac{75}{100} > \dfrac{26}{100}$; $\dfrac{3}{4} > 0.26$; so, $\dfrac{3}{4}$ is larger.

145.
$$\frac{3}{16} = \frac{12}{64}$$
$$+ \frac{9}{64} = \frac{9}{64}$$
$$\frac{21}{64}$$

147.
$$\frac{3}{5} = \frac{18}{30}$$
$$+ \frac{5}{6} = \frac{25}{30}$$
$$\frac{43}{30} = 1\frac{13}{30}$$

149.
$$2\frac{7}{16} = 2\frac{14}{32}$$
$$+ 6\frac{5}{32} = 6\frac{5}{32}$$
$$8\frac{19}{32}$$

151.
$$3\frac{7}{8} = 3\frac{28}{32}$$
$$5\frac{3}{16} = 5\frac{6}{32}$$
$$+ 1\frac{7}{32} = 1\frac{7}{32}$$
$$9\frac{41}{32} = 9 + 1\frac{9}{32} = 10\frac{9}{32}$$

153.
$$7\frac{5}{8} = 7\frac{10}{16}$$
$$+ 10\frac{7}{16} = 10\frac{7}{16}$$
$$17\frac{17}{16} = 17 + 1\frac{1}{16} = 18\frac{1}{16}$$

The pipe was $18\frac{1}{16}$ in. before it was cut.

155.
$$7\frac{5}{16} = 7\frac{10}{32}$$
$$+ 5\frac{9}{32} = 5\frac{9}{32}$$
$$12\frac{19}{32}$$

The original piece of stock was $12\frac{19}{32}$ in.

157.
$$3\frac{1}{8} = 3\frac{4}{32}$$
$$5\frac{3}{32} = 5\frac{3}{32}$$
$$+ 7\frac{9}{16} = 7\frac{18}{32}$$
$$15\frac{25}{32}$$

The welded rod is $15\frac{25}{32}$ in

159.
$$3\frac{1}{2} = 3\frac{2}{4}$$
$$+ 1\frac{1}{4} = 1\frac{1}{4}$$
$$4\frac{3}{4} \text{ in.}$$

161.
$$\frac{1}{2} = \frac{8}{16}$$
$$\frac{3}{16} = \frac{3}{16}$$
$$+ \frac{3}{16} = \frac{3}{16}$$
$$\frac{14}{16} = \frac{7}{8} \text{ in.}$$

163.
$$\frac{5}{9}$$
$$- \frac{2}{9}$$
$$\frac{3}{9} = \frac{1}{3}$$

165.
$$3\frac{5}{8} = 3\frac{5}{8}$$
$$- 2 = 2\frac{0}{8}$$
$$1\frac{5}{8}$$

167.
$$8\frac{7}{8} = 8\frac{28}{32} = 7\frac{32}{32} + \frac{28}{32} = 7\frac{60}{32}$$
$$- 2\frac{29}{32} = 2\frac{29}{32} = 2\frac{29}{32} \qquad = 2\frac{29}{32}$$
$$5\frac{31}{32}$$

169.
$$12\frac{11}{16}$$
$$- 5$$
$$7\frac{11}{16}$$

171.
$$122\frac{1}{2} = 121\frac{3}{2} = 121\frac{6}{4}$$
$$- 87\frac{3}{4} = -87\frac{3}{4} = -87\frac{3}{4}$$
$$34\frac{3}{4}$$

173.

$$\frac{7}{8} = \frac{14}{16}$$

$$\frac{3}{16} = \frac{3}{16} \qquad\qquad 2 = 1\frac{16}{16}$$

$$\frac{1}{16} = \frac{1}{16} \qquad - 1\frac{9}{16} = 1\frac{9}{16}$$

$$+\frac{7}{16} = \frac{7}{16} \qquad\qquad \frac{7}{16} \qquad \text{The metal is } \frac{7}{16} \text{ in. thick if it is flush with the bolt.}$$

$$\frac{25}{16} = 1\frac{9}{16}$$

175.

$$2\frac{5}{16} = 2\frac{20}{64} = 1\frac{84}{64}$$

$$-\frac{55}{64} = \frac{55}{64} = \frac{55}{64} \qquad \text{The difference in the diameters is } 1\frac{29}{64} \text{ in.}$$

$$1\frac{29}{64}$$

177. $\dfrac{1}{3} \times \dfrac{7}{8} = \dfrac{7}{24}$

179. $\dfrac{7}{9} \times \dfrac{3}{8} = \dfrac{7}{\cancel{9}_{3}} \times \dfrac{\cancel{3}^{1}}{8} = \dfrac{7}{24}$

181. $\dfrac{15}{16} \times \dfrac{4}{5} \times \dfrac{2}{3} = \dfrac{\cancel{15}^{1/3}}{\cancel{16}_{4/2}} \times \dfrac{\cancel{4}^{1}}{\cancel{5}} \times \dfrac{\cancel{2}^{1}}{\cancel{3}} = \dfrac{1}{2}$

183. $\dfrac{7}{16} \times 18 = \dfrac{7}{16} \times \dfrac{18}{1} = \dfrac{7}{\cancel{16}_{8}} \times \dfrac{\cancel{18}^{9}}{1} = \dfrac{63}{8} = 7\dfrac{7}{8}$

185. $1\dfrac{1}{2} \times \dfrac{4}{5} = \dfrac{3}{2} \times \dfrac{4}{5} = \dfrac{3}{\cancel{2}_{1}} \times \dfrac{\cancel{4}^{2}}{5} = \dfrac{6}{5} = 1\dfrac{1}{5}$

187. $1\dfrac{3}{4} \times 1\dfrac{1}{7} = \dfrac{\cancel{7}^{1}}{\cancel{4}_{1}} \times \dfrac{\cancel{8}^{2}}{\cancel{7}_{1}} = \dfrac{1 \times 2}{1 \times 1} = \dfrac{2}{1} = 2$

189. $(12 \times 8) + \left(12 \times \dfrac{3}{8}\right) = (12 \times 8) + \left(\dfrac{\cancel{12}^{3}}{1} \times \dfrac{3}{\cancel{8}_{2}}\right)$

$$= 96 + \dfrac{9}{2}$$

$$= 96 + 4\dfrac{1}{2} = 100\dfrac{1}{2}$$

The wall will be $100\dfrac{1}{2}$ in. high.

191. $3\dfrac{2}{3} \times \dfrac{3}{4} = \dfrac{11}{\cancel{3}_{1}} \times \dfrac{\cancel{3}^{1}}{4} = \dfrac{11}{4} = 2\dfrac{3}{4}$

The dessert used $2\dfrac{3}{4}$ cups of flour.

193. $18\dfrac{3}{4} \times 8$

$$\dfrac{75}{\cancel{4}_{1}} \times \dfrac{\cancel{8}^{2}}{1} = \dfrac{75 \times 2}{1 \times 1} = \dfrac{150}{1} = 150 \text{ cm}$$

195. $\left(\dfrac{3}{4}\right)^{2} = \dfrac{3}{4} \cdot \dfrac{3}{4} = \dfrac{9}{16}$

197. $\left(\dfrac{4}{9}\right)^2 = \dfrac{4}{9} \cdot \dfrac{4}{9} = \dfrac{16}{81}$

199. $\left(\dfrac{1}{2}\right)^3 = \dfrac{1}{2} \cdot \dfrac{1}{2} \cdot \dfrac{1}{2} = \dfrac{1}{8}$

201. $\left(\dfrac{9}{10}\right)^2 = \dfrac{9}{10} \cdot \dfrac{9}{10} = \dfrac{81}{100}$

203. $\dfrac{4}{1}$

Reciprocal $= \dfrac{1}{4}$

205. $0.7 = \dfrac{7}{10}$

Reciprocal $= \dfrac{10}{7}$ or $1\dfrac{3}{7}$

207. $\dfrac{7}{8} \div \dfrac{3}{4} = \dfrac{7}{8} \times \dfrac{4}{3} = \dfrac{7}{\overset{}{\underset{2}{8}}} \times \dfrac{\overset{1}{4}}{3} = \dfrac{7}{6} = 1\dfrac{1}{6}$

209. $\dfrac{7}{8} \div \dfrac{3}{32} = \dfrac{7}{8} \times \dfrac{32}{3} = \dfrac{7}{\underset{1}{8}} \times \dfrac{\overset{4}{32}}{3} = \dfrac{28}{3} = 9\dfrac{1}{3}$

211. $18 \div \dfrac{3}{4} = \dfrac{18}{1} \div \dfrac{3}{4} = \dfrac{\overset{6}{18}}{1} \times \dfrac{4}{\underset{1}{3}} = \dfrac{24}{1} = 24$

213. $5\dfrac{1}{10} \div 2\dfrac{11}{20} = \dfrac{51}{10} \div \dfrac{51}{20} = \dfrac{51}{10} \times \dfrac{20}{51} = \dfrac{\overset{1}{51}}{\underset{1}{10}} \times \dfrac{\overset{2}{20}}{\underset{1}{51}} = \dfrac{2}{1} = 2$

215. $7\dfrac{1}{5} \div 12 = \dfrac{36}{5} \div \dfrac{12}{1} = \dfrac{36}{5} \times \dfrac{1}{12} = \dfrac{\overset{3}{36}}{5} \times \dfrac{1}{\underset{1}{12}} = \dfrac{3}{5}$

217. $\dfrac{3}{16} \times 3 = \dfrac{3}{16} \times \dfrac{3}{1} = \dfrac{9}{16}$ Waste

$12 - \dfrac{9}{16} = 11\dfrac{16}{16} - \dfrac{9}{16} = 11\dfrac{7}{16}$

$11\dfrac{7}{16} \div 4 = \dfrac{183}{16} \div \dfrac{4}{1} = \dfrac{183}{16} \times \dfrac{1}{4} = \dfrac{183}{64} = 2\dfrac{55}{64}$

The maximum length that each pipe can be is $2\dfrac{55}{64}$ in.

219. $1\dfrac{1}{8} \div 6 = \dfrac{9}{8} \div \dfrac{6}{1} = \dfrac{9}{8} \times \dfrac{1}{6} = \dfrac{\overset{3}{9}}{8} \times \dfrac{1}{\underset{2}{6}} = \dfrac{3}{16}$

Each piece is $\dfrac{3}{16}$ yd long.

221. $22\dfrac{1}{2} \div \dfrac{5}{8} = \dfrac{45}{2} \div \dfrac{5}{8} = \dfrac{\overset{9}{45}}{\underset{1}{2}} \cdot \dfrac{\overset{4}{8}}{\underset{1}{5}} = \dfrac{9 \cdot 4}{1 \cdot 1} = 36$ lengths

223. $\dfrac{\frac{1}{3}}{6} = \dfrac{1}{3} \div 6 = \dfrac{1}{3} \div \dfrac{6}{1} = \dfrac{1}{3} \times \dfrac{1}{6} = \dfrac{1}{18}$

225. $\dfrac{8}{1\frac{1}{2}} = 8 \div 1\dfrac{1}{2} = \dfrac{8}{1} \div \dfrac{3}{2} = \dfrac{8}{1} \times \dfrac{2}{3} = \dfrac{16}{3} = 5\dfrac{1}{3}$

227. $\dfrac{2\frac{1}{5}}{8\frac{4}{5}} = 2\dfrac{1}{5} \div 8\dfrac{4}{5} = \dfrac{11}{5} \div \dfrac{44}{5} = \dfrac{11}{5} \times \dfrac{5}{44} = \dfrac{\overset{1}{11}}{\underset{1}{5}} \times \dfrac{\overset{1}{5}}{\underset{4}{44}} = \dfrac{1}{4}$

229. $\dfrac{12\frac{1}{2}}{100} = 12\dfrac{1}{2} \div 100 = \dfrac{25}{2} \div \dfrac{100}{1} = \dfrac{25}{2} \times \dfrac{1}{100} = \dfrac{\overset{1}{25}}{2} \times \dfrac{1}{\underset{4}{100}} = \dfrac{1}{8}$

231. $\dfrac{1 \text{ ft}}{12 \text{ in.}}$; $\dfrac{12 \text{ in.}}{1 \text{ ft}}$ **233.** $\dfrac{2{,}000 \text{ lb}}{1 \text{ T}}$; $\dfrac{1 \text{ T}}{2{,}000 \text{ lb}}$ **235.** 5 lb = _____ oz

$$\frac{5 \cancel{\text{lb}}}{1}\left(\frac{16 \text{ oz}}{1 \cancel{\text{lb}}}\right) = 80 \text{ oz}$$

237. 680 oz = _____ lb

$$\frac{680 \cancel{\text{oz}}}{1}\left(\frac{1 \text{ lb}}{16 \cancel{\text{oz}}}\right) = 42\frac{1}{2} \text{ lb}$$

239. $1\dfrac{1}{4}$ mi = _____ ft

$$1\frac{1}{4} \text{ mi} = \frac{5 \cancel{\text{mi}}}{4}\left(\frac{5{,}280 \text{ ft}}{1 \cancel{\text{mi}}}\right) = 6{,}600 \text{ ft}$$

To fence the property line, 6,600 ft of wire is needed.

241.   1 ft 19 in. =
1 ft + 1 ft 7 in. =
2 ft 7 in.

243.   12 lb $17\dfrac{1}{2}$ oz =

12 lb + 1 lb $1\dfrac{1}{2}$ oz =

13 lb $1\dfrac{1}{2}$ oz

245.   5 gal 3 qt
   + 2 gal 3 qt
    ‾‾‾‾‾‾‾‾‾‾
   7 gal 6 qt = 8 gal 2 qt

247.   4 lb 9 oz = 3 lb 16 oz + 9 oz = 3 lb 25 oz
   − 3 lb 11 oz   − 3 lb 11 oz
   ‾‾‾‾‾‾‾‾‾‾   ‾‾‾‾‾‾‾‾‾‾
                                    14 oz

249. 2 ft − 7 in. = 1 ft 12 in. − 7 in. = 1 ft 5 in. or 17 in.
The length of the water hose was 1 ft 5 in. after
it was cut.

251.   9 in.
   × 7 in.
   ‾‾‾‾‾‾
   63 in²

253. 20 yd 2 ft 6 in. ÷ 2
     10 yd 1 ft 3 in.
  2)‾20 yd 2 ft 6 in.
     20 yd 2 ft 6 in.

255. 18 lb ÷ 4

        4 lb   8 oz
   ‾‾‾‾‾‾‾‾‾‾
  4)‾18 lb
     16 lb
     ‾‾‾‾
      2 lb → 32 oz
          32 oz
          ‾‾‾‾‾

Each box would weigh 4 lb 8 oz.

257. 14 ft ÷ 4 ft = $\dfrac{14 \cancel{\text{ft}}}{4 \cancel{\text{ft}}} = 3\dfrac{1}{2}$ ft

259. $5 \dfrac{\text{mi}}{\text{min}}$ = _____ $\dfrac{\text{mi}}{\text{h}}$

$$5\frac{\text{mi}}{\cancel{\text{min}}}\left(\frac{60 \cancel{\text{min}}}{1 \text{ h}}\right) = 300 \frac{\text{mi}}{\text{h}}$$

261. 3 pt = _____ qt

$$\frac{3 \cancel{\text{pt}}}{1}\left(\frac{1 \text{ qt}}{2 \cancel{\text{pt}}}\right) = \frac{3}{2} \text{ qt} = 1\frac{1}{2} \text{ qt or 1 qt 1 pt}$$

The recipe calls for $1\dfrac{1}{2}$ quarts of milk.

Chapter 2 Practice Test

1. 3 out of 4 people in a survey, $\dfrac{3}{4}$

3. $\dfrac{9}{3} = 3\overline{)9}^{\,3} = 3$ To convert an improper fraction to a whole number, perform the division indicated.

5. $4\dfrac{6}{7} = \dfrac{(4 \times 7) + 6}{7} = \dfrac{34}{7}$ To convert a mixed number to an improper fraction, multiply the denominator of the fractional part by the whole number and add the numerator of the fractional part to form the new numerator, while the denominator stays the same.

7.
$$\dfrac{96}{}$$
$= 2 \cdot 48$
$= 2 \cdot 2 \cdot 24$
$= 2 \cdot 2 \cdot 2 \cdot 12$
$= 2 \cdot 2 \cdot 2 \cdot 2 \cdot 6$
$= 2 \cdot 2 \cdot 2 \cdot 2 \cdot 2 \cdot 3$
$2^5 \cdot 3$

9.
$$\dfrac{62}{}$$
$= 2 \cdot 31$

11.
$36 = 2^2(3^2)$
$45 = 3^2(5)$
$54 = 2(3^3)$
$\text{LCM} = 2^2(3^3)(5) = 540$

13.
$12 = 2^2(3)$
$18 = 2(3^2)$
$36 = 2^2(3^2)$
$\text{GCF} = 2(3) = 6$

15. $1\dfrac{1}{4}$ in.

17. $2\overline{)21.0}^{\,10.5}$; $\dfrac{21}{2} = 10.5$
$\dfrac{2}{}$
$\dfrac{10}{}$
10

19.
$\dfrac{4}{5} = \dfrac{8}{10}$
$\dfrac{7}{10} = \dfrac{7}{10}$
$\dfrac{7}{10}$ is smaller because $7 < 8$.

21.
$3\dfrac{4}{15} = 3\dfrac{8}{30}$
$4\dfrac{3}{10} = 4\dfrac{9}{30}$
$\overline{\phantom{4\dfrac{3}{10}}}$
$7\dfrac{17}{30}$

23. $2\dfrac{2}{9} \times 1\dfrac{3}{4} =$

$\dfrac{\overset{5}{\cancel{20}}}{9} \times \dfrac{7}{\underset{1}{\cancel{4}}} = \dfrac{35}{9} = 3\dfrac{8}{9}$

25.
$6\dfrac{1}{4} = 5\dfrac{4}{4} + \dfrac{1}{4} = 5\dfrac{5}{4}$
$-2\dfrac{3}{4} = 2\dfrac{3}{4} \phantom{+\dfrac{1}{4}} = 2\dfrac{3}{4}$
$\overline{\phantom{-2\dfrac{3}{4} = 2\dfrac{3}{4}}}$
$3\dfrac{2}{4} = 3\dfrac{1}{2}$

27. $7\dfrac{1}{2} \div \dfrac{5}{9} = \dfrac{15}{2} \div \dfrac{5}{9} = \dfrac{\overset{3}{\cancel{15}}}{2} \cdot \dfrac{9}{\underset{1}{\cancel{5}}} = \dfrac{27}{2} = 13\dfrac{1}{2}$

29. $\dfrac{2}{7}$ received safety awards.

31.
$5\dfrac{1}{2} = 5\dfrac{3}{6} = 4\dfrac{6}{6} + \dfrac{3}{6} = 4\dfrac{9}{6}$
$-1\dfrac{2}{3} = 1\dfrac{4}{6} = 1\dfrac{4}{6} \phantom{+\dfrac{3}{6}} = 1\dfrac{4}{6}$
$\overline{\phantom{-1\dfrac{2}{3} = 1\dfrac{4}{6} = 1\dfrac{4}{6}}}$
$3\dfrac{5}{6}$

There were $3\dfrac{5}{6}$ cups of sugar left.

33. $2\dfrac{2}{3} \cdot 3 = \dfrac{(2 \times 3) + 2}{3} \cdot \dfrac{3}{1} = \dfrac{8}{3} \cdot \dfrac{3}{1} = \dfrac{8}{\underset{1}{\cancel{3}}} \cdot \dfrac{\overset{1}{\cancel{3}}}{1} = \dfrac{8}{1} = 8$

To make three costumes, 8 yards of red satin material would be needed.

Chapter Review Exercises

1. $\dfrac{72}{100} = 72\%$ **3.** $\dfrac{23}{100} = 23\%$ **5.** $0.7 = 70\%$

7. $0.83 = 83\%$ **9.** $3\dfrac{1}{5} = 3.20 = 320\%$ **11.** $125 = 12{,}500\%$

13. $17.3 = 1{,}730\%$ **15.** $72\% = 0.72 = \dfrac{72}{100} = \dfrac{18}{25}$

17. $12\dfrac{1}{2}\% = \dfrac{12\frac{1}{2}\%}{100\%} = \dfrac{\frac{25}{2}}{100} = \dfrac{25}{2} \cdot \dfrac{1}{100} = \dfrac{25}{200} = \dfrac{1}{8}$

19. $\dfrac{2}{3}\% = \dfrac{\frac{2}{3}}{100} = \dfrac{2}{3} \cdot \dfrac{1}{100} = \dfrac{2}{300} = \dfrac{1}{150}$

21. $275\% = \dfrac{275\%}{100\%} = \dfrac{275}{100} = 2\dfrac{75}{100} = 2\dfrac{3}{4}$

23. $112\dfrac{1}{2}\% = \dfrac{112\frac{1}{2}\%}{100\%} = \dfrac{225}{2} \cdot \dfrac{1}{100} = \dfrac{225}{200} = 1\dfrac{25}{200} = 1\dfrac{1}{8}$

25. $227.2\% = 2.272$ **27.** $340\% = 3.4$ **29.** $83\% = 0.83$ **31.** $62\dfrac{1}{2}\% = 62.5\% = 0.625$

33. $\dfrac{1}{2} = \dfrac{a}{9}$; $1(9) = 9$; $2\overline{)9.0}$

$$\begin{array}{r} 4.5 \\ \hline 9.0 \\ 8 \\ \hline 10 \\ 10 \\ \hline \end{array}$$

35. $\dfrac{7}{16} = \dfrac{21}{y}$

$y = \dfrac{21(16)}{7}$

$y = \dfrac{336}{7}$

$y = 48$

37.
	R	B	P
	5%	of 180	is what number?
	percent	total	portion

39.
	R	B	P
	45 percent of	how many dollars is	$36?
	percent	total	portion

41.
	P	R	B
6 is what	percent of	25	sacks of grass seed?
portion	percent	total	

43.
$$\overset{R}{18\%} \quad \text{of} \quad \overset{B}{150} \quad \overset{P}{\text{pieces of sod is how many?}}$$
$$\underset{\text{percent}}{} \quad \underset{\text{base}}{} \quad \underset{\text{portion}}{}$$

45. 5% "of" 480
$$\overset{R}{} \qquad \overset{B}{}$$
$$P = RB$$
$$P = 0.05(480)$$
$$P = 24$$

47. $\dfrac{1}{4}\%$ "of" 175
$$\underset{R}{} \qquad \underset{B}{}$$
$\dfrac{1}{4}\% = 0.25\% = 0.0025$
$$P = RB$$
$$P = 0.0025(175)$$
$$P = 0.4375$$

49. 39 "is" "of" 65
$$\underset{R}{} \qquad \underset{B}{}$$
$R = \dfrac{P}{B}$
$$R = \dfrac{39}{65}(100\%)$$
$$R = 0.6(100\%)$$
$$R = 60\%$$

51. "of" 65 "is" 162.5
$$\underset{B}{} \qquad \underset{R}{}$$
$R = \dfrac{162.5}{65}(100\%)$
$$R = 2.5(100\%)$$
$$R = 250\%$$

53. 24% "is" 19.92
$$\underset{R}{} \quad \underset{P}{}$$
$\dfrac{R}{100} = \dfrac{P}{B}$
$$\dfrac{24}{100} = \dfrac{19.92}{B}$$
$$24B = 100 \times 19.92$$
$$24B = 1{,}992$$
$$B = \dfrac{1{,}992}{24}$$
$$B = 83$$

55. 260% "is" 395.2
$$\underset{R}{} \qquad \underset{P}{}$$
$\dfrac{R}{100} = \dfrac{P}{B}$
$$\dfrac{260}{100} = \dfrac{395.2}{B}$$
$$260B = 100(395.2)$$
$$260B = 39{,}520$$
$$260B = \dfrac{39{,}520}{260}$$
$$B = 152$$

57. 38.25 "is" "of" 250
$$\underset{P}{} \qquad \underset{B}{}$$
$\dfrac{R}{100} = \dfrac{P}{B}$
$$\dfrac{R}{100} = \dfrac{38.25}{250}$$
$$250R = 100(38.25)$$
$$250R = 3{,}825$$
$$250R = \dfrac{3{,}825}{250}$$
$$R = 15.3\%$$

59. "of" 26 "is" 130
$$\underset{B}{} \qquad \underset{P}{}$$
$\dfrac{R}{100} = \dfrac{P}{B}$
$$\dfrac{R}{100} = \dfrac{130}{26}$$
$$26R = 100(130)$$
$$26R = 13{,}000$$
$$R = \dfrac{13{,}000}{26}$$
$$R = 500\%$$

61. $10\dfrac{1}{3}\%$ "is" 8.68
$$\underset{R}{} \qquad \underset{P}{}$$
$\dfrac{R}{100} = \dfrac{P}{B}$
$$\dfrac{10\frac{1}{3}}{100} = \dfrac{8.68}{B}$$
$$10\dfrac{1}{3} \times B = 100 \times 8.68$$
$$10\dfrac{1}{3} \times B = 868$$
$$B = 868 \div 10\dfrac{1}{3}$$
$$B = \dfrac{868}{1} \div \dfrac{31}{3}$$
$$B = \dfrac{868}{1} \cdot \dfrac{3}{31}$$
$$B = 28(3) = 84$$

63. 84 part 224 total

$\dfrac{R}{100} = \dfrac{P}{B}$

$\dfrac{R}{100} = \dfrac{84}{224}$

$R = \dfrac{100(84)}{224}$

$R = 37.5\%$

The alloy is 37.5% zinc.

65. 27 part 2374 total

(nearest hundreth percent)

$\dfrac{R}{100} = \dfrac{P}{B}$

$\dfrac{R}{100} = \dfrac{27}{2,374}$

$R = \dfrac{100(27)}{2,374}$

$R = 1.137320977\%$

$R = 1.14\%$

There was 1.14% defective.

67. 28% 950 total

$\dfrac{R}{100} = \dfrac{P}{B}$

$\dfrac{28}{100} = \dfrac{P}{950}$

$P = \dfrac{28(950)}{100}$

$P = 266$

Of the paycheck, $266 goes for food.

69. 1475 part 36,875 total

$\dfrac{R}{100} = \dfrac{P}{B}$

$\dfrac{R}{100} = \dfrac{1475}{36,875}$

$R = \dfrac{100(1475)}{36,875}$

$R = 4\%$

Of those receiving traffic citations, 4% were prosecuted.

71. 67 part 33.5%

$\dfrac{R}{100} = \dfrac{P}{B}$

$\dfrac{33.5}{100} = \dfrac{67}{B}$

$B = \dfrac{100(67)}{33.5}$

$B = 200$

The total number of students was 200.

73. $3,296 total 6.2%

$\dfrac{R}{100} = \dfrac{P}{B}$

$\dfrac{6.2}{100} = \dfrac{P}{3,296}$

$P = \dfrac{6.2(3,296)}{100}$

$P = 204.352$

The social security tax is $204.35

75.

$$\begin{array}{r} 4,873 \text{ all} \\ -\,2,500 \\ \hline 2,373 \text{ base for} \\ \text{commission} \end{array}$$

$\dfrac{8}{100} = \dfrac{P}{2,373}$

$P = \dfrac{8(2,373)}{100}$

$P = 189.84$

$$\begin{array}{r} 189.84 \\ +\,250 \\ \hline 439.84 \end{array}$$

The distributor's salary is $439.84 for the given week.

77.

$$\begin{array}{r} 100\% \text{ all} \\ +\;\;5\% \text{ tax} \\ \hline 105\% \end{array}$$

$\dfrac{105}{100} = \dfrac{P}{348.25}$

$P = \dfrac{105(348.25)}{100}$

$P = 365.6625$

The total bill is $365.66.

79. $\dfrac{6.2}{100} = \dfrac{P}{27{,}542}$

$P = \dfrac{6.2(27{,}542)}{100}$

$P = 1{,}707.604$

The employer's contributions were $1,707.60.

81. $\dfrac{R}{100} = \dfrac{4.96}{283.15}$

$R = \dfrac{100(4.96)}{283.15}$

$R = 1.751721702\%$

The monthly interest rate is 1.75% on the account.

83. $\dfrac{5}{100} = \dfrac{P}{52{,}475}$ $\left(18 \text{ month} = \dfrac{18}{12} \text{ year} = 1.5 \text{ years}\right)$

$P = \dfrac{5(52{,}475)}{100}$ $1.5 \times 2{,}623.75 = 3{,}935.625$

$P = 2{,}623.75$

There was $3,935.63 interest earned.

85. $\dfrac{3.25}{100} = \dfrac{P}{27.45}$

$P = \dfrac{3.25(27.45)}{100}$

$P = 0.892125$

There is $0.89 sales tax on the purchase.

87. $\dfrac{R}{100} = \dfrac{170}{2{,}000}$

$R = \dfrac{100(170)}{2{,}000}$

$R = 8.5\%$

The interest rate was 8.5%.

89. $\dfrac{9.75}{100} = \dfrac{P}{6{,}500}$ $\left(7 \text{ month} = \dfrac{7}{12} \text{ year}\right)$

$P = \dfrac{9.75(6{,}500)}{100}$ $\dfrac{7}{12} \times 633.75 = 369.6875$

$P = 633.75$

The interest was $369.69 on the business loan after 7 months.

91. $\dfrac{3}{100} = \dfrac{10.65}{B}$

$B = \dfrac{100(10.65)}{3}$

$B = 355$

The sales were $355 for the weekend.

93. $100\% - 12.9\% = 87.1\%$

95. $100\% - 21.5\% = 78.5\%$ **97.** $100\% - 32\dfrac{2}{5}\% = 67\dfrac{3}{5}\%$ **99.** $100\% - 92\% = 8\%$

101. $\begin{array}{r} 800 \\ -\ 750 \\ \hline 50 \end{array}$ end, original, required, B increase, P $\dfrac{R}{100} = \dfrac{P}{B}$

$\dfrac{R}{100} = \dfrac{50}{750}$

$R = \dfrac{100(50)}{750}$

$R = 6\dfrac{2}{3}\% \text{ or } 7\%$

There was 7% ordered for waste.

103.   600   original, B

$\underline{-\ 516}$ end

84 decrease, P

$$\frac{R}{100} = \frac{P}{B}$$

$$\frac{R}{100} = \frac{84}{600}$$

$$R = \frac{100(84)}{600}$$

$$R = 14\%$$

The percent of decrease in the price of the lathe is 14%.

105.   100%   all

$\underline{-\ \ 2\%}$ loss

98%

$$\frac{R}{100} = \frac{P}{B}$$

$$\frac{98}{100} = \frac{P}{145}$$

$$P = \frac{98(145)}{100}$$

$$P = 142.1$$

The dried casting will weigh 142.1 kg.

107. $$\frac{R}{100} = \frac{P}{B}$$

$$\frac{0.8}{100} = \frac{P}{62.5}$$

$$P = \frac{0.8(62.5)}{100}$$

$$P = 0.5$$

62.5	62.5
$-\ 0.5$	$+\ 0.5$
62.0	63.0

The limit dimensions of the length of the part are 62 cm to 63 cm.

109.   100%   all

$\underline{-25\%}$ less

75%

$$\frac{R}{100} = \frac{P}{B}$$

$$\frac{75}{100} = \frac{P}{2.25}$$

$$P = \frac{75(2.25)}{100}$$

$$P = 1.6875$$

(assume rounding to the nearest cent)
The computer disk now sells for $1.69.

111. $$\frac{R}{100} = \frac{300}{2,400}$$

$$R = \frac{100(300)}{2,400}$$

$$R = 12.5\%$$

The percent of increase is 12.5%.

113.   100%   all

$+15\%$ more

115%

$$\frac{R}{100} = \frac{P}{B}$$

$$\frac{115}{100} = \frac{P}{50}$$

$$P = \frac{115(50)}{100}$$

$$P = 57.5$$

The larger carton will hold 57.5 pounds.

115.   24   end

$\underline{-\ 18}$ original, B

6 increase, P

$$\frac{R}{100} = \frac{P}{B}$$

$$\frac{R}{100} = \frac{6}{18}$$

$$R = \frac{100(6)}{18}$$

$$R = 33\frac{1}{3}\%$$

The percent of increase was $33\frac{1}{3}\%$.

117.   168   original, B

$\underline{-\ 160}$ end

8 loss, P

$$\frac{R}{100} = \frac{P}{B}$$

$$\frac{R}{100} = \frac{8}{168}$$

$$R = \frac{100(8)}{168}$$

$$R = 4.761904762\%$$

The percent of weight loss was about 4.8%.

119.
$$\begin{array}{r} 100\ \%\ \text{all} \\ -\ 17.4\%\ \text{less} \\ \hline 82.6\% \end{array}$$
(nearest whole number)

$$\frac{R}{100} = \frac{P}{B}$$

$$\frac{82.6}{100} = \frac{P}{350}$$

$$P = \frac{82.6(350)}{100}$$

$$P = 289.1 \approx 289$$

The horsepower of the new car is about 289.

121.
$$\begin{array}{r} 3{,}900\ \text{jobs in 2010} \\ -\ 2{,}095\ \text{jobs in 2000} \\ \hline 1{,}805\ \text{increase} \end{array}$$

$$\frac{R}{100} = \frac{P}{B}$$

$$\frac{R}{100} = \frac{1{,}805}{2{,}095}$$

$$R = \frac{100(1{,}805)}{2{,}095}$$

$$R = 86.1575179\%$$

The percent increase is 86%.

123.
$$\begin{array}{r} 3{,}420 \\ 652 \\ 625 \\ +\ 150 \\ \hline 4{,}847 \end{array}$$

$$\frac{R}{100} = \frac{4{,}847}{18{,}250}$$

$$R = \frac{100(4{,}847)}{18{,}250}$$

$$R = 26.55890411\%$$

Of the motorist's annual income, 26.6% is used for automotive transportation.

125.
$$\begin{array}{r} 35\ \text{total} \\ -\ 25\ \text{women} \\ \hline 10\ \text{men} \end{array}$$

$$\frac{R}{100} = \frac{10}{35}$$

$$R = \frac{100(10)}{35}$$

$$R = 28.57142857\%$$

The class had 28.6% men.

Chapter 3 Practice Test

1. $\dfrac{4}{5} = 0.8(100\%) = 80\%$

3. $0.3\% \div 100\% = 0.003$

5. $\underset{\text{percent}}{40\%}$ of $\underset{\text{total}}{10}$ x-ray techiques is how many? $\underset{\text{portion}}{}$

$$\begin{array}{lll} & & R \qquad\qquad B \qquad\qquad\qquad\qquad\qquad P \end{array}$$

7. $\underset{\text{portion}}{9}$ is what $\underset{\text{percent}}{\text{percent}}$ of $\underset{\text{total}}{27}$ dogwood trees?

$$\begin{array}{lll} P \qquad\qquad R \qquad\qquad B \end{array}$$

9. $\underset{\text{percent}}{12\%}$ of $\underset{\text{total}}{50}$ grass plugs is how many? $\underset{\text{portion}}{}$

$$\begin{array}{lll} R \qquad\qquad B \qquad\qquad\qquad P \end{array}$$

11. $6\frac{1}{4}\%$ "of" 144
$$ R \qquad\qquad B$$

$$\frac{R}{100} = \frac{P}{B}$$

$$\frac{6\frac{1}{4}}{100} = \frac{P}{144}$$

$$\frac{6\frac{1}{4}(144)}{100} = P$$

$$\frac{900}{100} = P$$

$$9 = P$$

13. 45.75 "is" 15%
$$ P \qquad\qquad R$$

$$\frac{R}{100} = \frac{P}{B}$$

$$\frac{15}{100} = \frac{45.75}{B}$$

$$B = \frac{100(45.75)}{15}$$

$$B = \frac{4{,}575}{15}$$

$$B = 305$$

15. 250% "is" 287.5
    R   P

$$\frac{R}{100} = \frac{P}{B}$$

$$\frac{250}{100} = \frac{287.5}{B}$$

$$B = \frac{100(287.5)}{250}$$

$$B = \frac{28,750}{250}$$

$$B = 115$$

17. 245% "is" 164.4
    R   P

nearest hundredth

$$\frac{R}{100} = \frac{P}{B}$$

$$\frac{245}{100} = \frac{164.4}{B}$$

$$B = \frac{100(164.4)}{245}$$

$$B = \frac{16,440}{245}$$

$$B = 67.10204082$$

$$B = 67.10$$

19. $100\% - 88\% = 12\%$

21. $15,000 total 14%
    B   P

$$\frac{R}{100} = \frac{P}{B}$$

$$\frac{14}{100} = \frac{P}{15,000}$$

$$14(15,000) = 100P$$

$$\frac{210,000}{100} = P$$

$$2,100 = P$$

3 months = $\frac{3}{12}$ year

$$\frac{3}{12}(2,100) =$$

$$\frac{1}{4}(2,100) =$$

$$525$$

There is $525 interest earned on the investment.

23.
$$\frac{R}{100} = \frac{P}{B}$$

$$\frac{7}{100} = \frac{175}{B}$$

$$B = \frac{100(175)}{7}$$

$$B = \frac{175,000}{7}$$

$$B = 2,500$$

The sales person sold $2,500.

25.
  149 end
 -123 original, B
   26 increase, P
(nearest hundredth)

$$\frac{R}{100} = \frac{P}{B}$$

$$\frac{R}{100} = \frac{26}{123}$$

$$R = \frac{100(26)}{123}$$

$$R = 21.13821138\%$$

The percent increase of women students was 21.14%.

27.
  372 original, B
 -323 end
   49 decrease, P
(nearest hundredth)

$$\frac{R}{100} = \frac{P}{B}$$

$$\frac{R}{100} = \frac{49}{372}$$

$$R = \frac{100(49)}{372}$$

$$R = 13.172034301\%$$

The percent of decrease was 13.17%.

29.
  36.6 original, B
 -34.7 end
   1.9 loss, P
(nearest whole number)

$$\frac{R}{100} = \frac{P}{B}$$

$$\frac{R}{100} = \frac{1.9}{36.6}$$

$$R = \frac{100(1.9)}{36.6}$$

$$R = 5.191256831\%$$

The percent of weight loss is about 5%.

31.
$$\frac{R}{100} = \frac{P}{B}$$

$$\frac{12}{100} = \frac{P}{873.92}$$

$$P = \frac{12(873.92)}{100}$$

$$P = 104.8704$$

The discount was $104.87

Chapters 1–3 Cumulative Practice Test

1. $12 + 5(4) \div 2 - 8$
$= 12 + 20 \div 2 - 8$
$= 12 + 10 - 8$
$= 22 - 8$
$= 14$

3. $P = 2l + 2w$
$P = 2(40 \text{ cm}) + 2(24 \text{ cm})$
$P = 80 \text{ cm} + 48 \text{ cm}$
$P = 128 \text{ cm}$

5. 42.8<u>1</u>96   1 is in the hundredths place so examine the digit to its right, 9. Since 9 is 5 or more, round up. Keep all digits on the left of 1 and drop digits on its right.

The rounded result is 42.82.

7. $420 = 2(210)$
$ = 2(2)(105)$
$ = 2(2)(3)(35)$
$ = 2(2)(3)(5)(7)$
Factored form $= 2(2)(3)(5)(7)$
Exponential notation $= 2^2(3)(5)(7)$

9. 42.8
23.06
15.9
$\overline{81.76}$

11. 52.06
8.723
$\overline{15618}$
10412
36442
41648
$\overline{454.11938}$

13. $3\dfrac{7}{8} = 3\dfrac{7}{8}$
$9\dfrac{3}{4} = 9\dfrac{6}{8}$
$5\dfrac{1}{2} = 5\dfrac{4}{8}$
$\overline{}$
$17\dfrac{17}{8} = 17 + 2\dfrac{1}{8} = 19\dfrac{1}{8}$

15. $4\dfrac{1}{2} \times \dfrac{4}{9} = \dfrac{9}{2} \times \dfrac{4}{9} = 2$

17. $1\dfrac{3}{4} \div 1\dfrac{5}{12} = \dfrac{7}{4} \div \dfrac{17}{12} = \dfrac{7}{4} \times \dfrac{12}{17} = \dfrac{21}{17} = 1\dfrac{4}{17}$

19. $3.5^2 = 3.5(3.5) = 12.25$

21. $\sqrt{1.96} = 1.4$

23. $\sqrt{3,600} = \sqrt{36 \cdot 100} = 6 \cdot 10 = 60$

25. $125 \text{ in.}\left(\dfrac{1 \text{ ft}}{12 \text{ in.}}\right) = 10\dfrac{5}{12} \text{ ft or } 10 \text{ ft } 5 \text{ in.}$

27. $147 \text{ ft } 11 \text{ in.} =$
$147 \text{ ft} + 11 \text{ in.} =$
$147 \text{ ft}\left(\dfrac{12 \text{ in.}}{1 \text{ ft}}\right) + 11 \text{ in.} =$
$1,764 \text{ in.} + 11 \text{ in.} =$
$1,775 \text{ in.}$

29. What percent of 24 is 20?
$R = \dfrac{P}{B}(100\%)$
$R = \dfrac{20}{24}(100\%)$
$R = 83\%$ (rounded)

31. Amount of increase $= \$45,000 - \$42,000$
$ = \$3,000$
Percent of increase $= \dfrac{\$3,000}{\$42,000} \times 100\% = 7.14\%$ (rounded)

33. Complement of 14%: $100\% - 14\% = 86\%$

35. Complement of 82.0%: $100\% - 82\% = 18\%$
$8.0\% \div 18\% = 0.08 \div 0.18 = 0.444444$
DM basis percent of protein $= 0.444444 \times 100\% = 44.4\%$ (rounded)

chapter 4 Measurement

Chapter Review Exercises

1. 1,000 times
kilo -

3. $\dfrac{1}{1,000}$ of
milli -

5. $\dfrac{1}{100}$ of
centi -

7. dekameter (dkm)
deka - means 10 times.

9. milligram (mg)
milli - means $\dfrac{1}{1,000}$ of.

11. kiloliter (kL)
kilo - means 1,000 times.

13. Height of the Washington Monument
(a)   200 m

15. Weight of an egg
(a)   50 g

17. Weight of a man's shoe
(c)   0.25 kg

19. Bottle of medicine
(b)   50 mL

21. 61.7 m = _____ dkm

From m to dkm, move one space to the left.
67.1 m = 67.1 = 6.71 dkm

23. 2.3 m = _____ mm

From m to mm, move 3 spaces to the right.
2.3 m = 2.300 = 2,300 mm

25. 0.123 hm = _____ mm

From hm to mm, move 5 spaces to the right.
0.123 hm = 0.12300 = 12,300 mm

27. 23 dkm = _____ mm

From dkm to mm, move 4 spaces to the right.
23 dkm = 23.0000 = 230,000 mm

29. 41,327 dkm = _____ km

From dkm to km, move 2 spaces to the left.
41,327 dkm = 41327. = 413.27 km

31. 394.5 g = _____ hg

From g to hg, move 2 spaces to the left.
394.5 g = 394.5 = 3.945 hg

33. 3,000,974 cg = _____ kg

From cg to kg, move 5 spaces to the left.
3,000,974 cg = 3000974. = 30.00974 kg

35. 12 g + 5 m
Since g (grams) is a measure of weight
and m (meters) is a measure of length,
these cannot be added.

37. 8 g – 52 cg
800 cg – 52 cg = 748 cg
or
8 g – 0.52 g = 7.48 g

39. 6.83 cg × 9 = 61.47 cg

27

41. $7.5 \text{ kg} \div 0.5 \text{ kg} = \dfrac{7.5 \text{ k\!\!\!/g}}{0.5 \text{ k\!\!\!/g}} = 15$

43. $34 \text{ hL} \div 4 = 8\dfrac{1}{2} \text{ hL or } 8.5 \text{ hL}$

45. $2.7 \text{ m} \times 7 = 18.9 \text{ m}$
18.9 m of fabric must be purchased.

47. 5 mL + 24 cL
5 mL + 240 mL = 245 mL
or
0.5 cL + 24 cL = 24.5 cL
The recipe calls for 245 mL or 24.5 cL of liquid.

49. $42 \text{ m} \div 7 = 6 \text{ m}$
Each piece is 6 m long.

51. $25 \text{ L} \div 25 \text{ cL} = \dfrac{25 \text{ L}}{25 \text{ cL}} = \dfrac{2500 \text{ c\!\!\!/L}}{25 \text{ c\!\!\!/L}} = 100$

There are 100 servings of punch.

53. $\dfrac{72 \text{ h}}{1} \left(\dfrac{1 \text{ day}}{24 \text{ h}} \right) = 3 \text{ days}$

55. $\dfrac{158 \text{ min}}{1} \left(\dfrac{1 \text{ h}}{60 \text{ min}} \right) = 2 \text{ h } 38 \text{ min or } 2.63 \text{ h}$

57. $\dfrac{96 \text{ h}}{1} \left(\dfrac{1 \text{ day}}{24 \text{ h}} \right) = 4 \text{ days}$

59. $\dfrac{39 \text{ mo}}{1} \left(\dfrac{1 \text{ yr}}{12 \text{ mo}} \right) = 3 \text{ yr } 3 \text{ mo or } 3.25 \text{ yr}$

61. $15^\circ\text{C} = \underline{\quad\quad}\ ^\circ\text{F}$

$^\circ\text{F} = \dfrac{9}{5}\,^\circ\text{C} + 32$

$^\circ\text{F} = \dfrac{9}{5}\,(15) + 32$

$^\circ\text{F} = 27 + 32$

$^\circ\text{F} = 59$

63. $95^\circ\text{C} = \underline{\quad\quad}\ ^\circ\text{F}$

$^\circ\text{F} = \dfrac{9}{5}\,^\circ\text{C} + 32$

$^\circ\text{F} = \dfrac{9}{5}\,(95) + 32$

$^\circ\text{F} = 171 + 32$

$^\circ\text{F} = 203$

65. $40^\circ\text{C} = \underline{\quad\quad}\ ^\circ\text{F}$

$^\circ\text{F} = \dfrac{9}{5}\,^\circ\text{C} + 32$

$^\circ\text{F} = \dfrac{9}{5}\,(40) + 32$

$^\circ\text{F} = 72 + 32$

$^\circ\text{F} = 104$

67. $365^\circ\text{F} = \underline{\quad\quad}\ ^\circ\text{C}$

$^\circ\text{C} = \dfrac{5}{9}\,(^\circ\text{F} - 32)$

$^\circ\text{C} = \dfrac{5}{9}\,(365 - 32)$

$^\circ\text{C} = \dfrac{5}{9}\,(333)$

$^\circ\text{C} = 185$

69. $37.9^\circ\text{C} = \underline{\quad\quad}\ ^\circ\text{F}$

$^\circ\text{F} = \dfrac{9}{5}\,^\circ\text{C} + 32$

$^\circ\text{F} = \dfrac{9}{5}\,(37.9) + 32$

$^\circ\text{F} = 68.22 + 32$

$^\circ\text{F} = 100.2 \ \text{(rounded)}$

71. $215 \text{ m} = \underline{\quad\quad} \text{ yards}$

$\dfrac{215 \text{ m\!\!\!/}}{1} \left(\dfrac{1.0936 \text{ yd}}{1 \text{ m\!\!\!/}} \right) = 235.124 \text{ yd}$

73. $15 \text{ L} = \underline{\quad\quad} \text{ quarts}$

$\dfrac{15 \text{ L\!\!\!/}}{1} \left(\dfrac{1.0567 \text{ qt}}{1 \text{ L\!\!\!/}} \right) = 15.8505 \text{ qt}$

75. $32 \text{ kg} = \underline{\quad\quad} \text{ pounds}$

$\dfrac{32 \text{ k\!\!\!/g}}{1} \left(\dfrac{2.2046 \text{ lb}}{1 \text{ k\!\!\!/g}} \right) = 70.5472 \text{ lb}$

77. $9 \text{ in.} = \underline{\quad\quad} \text{ centimeters}$

$\dfrac{9 \text{ i\!\!\!/n.}}{1} \left(\dfrac{2.54 \text{ cm}}{1 \text{ i\!\!\!/n.}} \right) = 22.86 \text{ cm}$

79. $14.8 \text{ dkL} = \underline{\quad\quad} \text{ quarts}$

14.8 dkL = 148 L

$\dfrac{148 \text{ L\!\!\!/}}{1} \left(\dfrac{1.0567 \text{ qt}}{1 \text{ L\!\!\!/}} \right) = 156.3916 \text{ qt}$

81. 200 ft = _____ m

$$\frac{200 \text{ ft}}{1} \left(\frac{0.3048 \text{ m}}{1 \text{ ft}} \right) = 60.96 \text{ m}$$

83. 175 mi = _____ km

$$\frac{175 \text{ mi}}{1} \left(\frac{1.6093 \text{ km}}{1 \text{ mi}} \right) = 281.6275 \text{ km}$$

85. $142 \times 0.4536 = 64.4112$ Change lb to kg using conversion factor.

5 ft 9 in. = 5(12 + 9)   Change ft to in. using conversion factor.

$\qquad = 60 \text{ in.} + 9 \text{ in.}$

$\qquad = 69 \text{ in.}$

$69 \times 0.0254 = 1.7526$ Change in. to meters using conversion factor.

$$\text{BMI} = \frac{w}{h^2}$$

$$= \frac{64.4112}{(1.7526)^2}$$

$$= \frac{64.4112}{3.07160676}$$

$$= 20.96987181$$

$$\text{BMI} = 21$$

87. 304,243   6 significant digits

89. 4.010   4 significant digits

91. $2\frac{1}{2}$ in.   Precision $= \frac{1}{2}$ in.

$\frac{1}{2} \times \frac{1}{2} = \frac{1}{4}$ in. Greatest possible error

93. $3\frac{5}{16}$ ft   Precision $= \frac{1}{16}$ ft

$\frac{1}{2} \times \frac{1}{16} = \frac{1}{32}$ ft Greatest possible error

95. 5.8 cm   Precision = 0.1 cm

$0.5 \times 0.1 = 0.05$ cm Greatest possible error

97. 15.3 cm   Precision = 0.1 cm

$0.5 \times 0.1 = 0.05$ cm Greatest possible error

99. $|5.00 - 4.95| = 0.05$ gal absolute error

$\frac{0.05}{5.00} = 0.01$ relative error

$0.01 \times 100\% = 1\%$ percent error

101. 0.5010   4 significant digits

The zero between 5 and 1 is a significant digit and the zero on the right is a significant digit.

103. Precision = 0.1 cg

Greatest possible error $= \frac{1}{2} \times 0.1 = 0.05$ cg

105. 4.2 m + 5.08 m + 31.72 m + 5.46 m = 46.46 m or 46.5 m (using appropriate precision)

or 420 cm + 508 cm + 3,172 cm + 546 cm = 4,650 cm

107. 117 mm or 118 mm   **109.** 99 mm   **111.** 60 mm   **113.** 45 mm   **115.** 20 mm

117. $12\dfrac{7}{8}$ in.

$-\ 7\dfrac{5}{8}$ in.

$5\dfrac{2}{8}$ in. $= 5\dfrac{1}{4}$ in.

119. 8 in. $=\ 7\dfrac{32}{32}$ in.

$-\ 3\dfrac{5}{32}$ in. $=-\ 3\dfrac{5}{32}$ in.

$4\dfrac{27}{32}$ in.

121. 20.5 cm

$-\ 17.8$ cm

2.7 cm

123. $15\dfrac{3}{8}$ in.

$+\ 25\dfrac{1}{8}$ in.

$40\dfrac{4}{8} = 40\dfrac{1}{2}$ in.

$40\dfrac{1}{2} \div 2 = \dfrac{81}{2} \cdot \dfrac{1}{2} = \dfrac{81}{4} = 20\dfrac{1}{4}$ in.

125. 8.9 cm

$+2.3$ cm

11.2 cm

$11.2 \div 2 = 5.6$ cm

127. 9.8 cm

$+5\ \ $ cm

14.8 cm

14.8 cm $\div\ 2 = 7.4$ cm

Chapter 4 Practice Test

1. 298 m $=$ _____ km

From m to km, move 3 digits to the left.

298 m $= \underset{\sim}{298.} = 0.298$ km

3. 10 L $- 5.2$ dL

100 dL $- 5.2$ dL $= 94.8$ dL

or

10 L $- 0.52$ L $= 9.48$ L

There are 94.8 dL or 9.48 L of liquid remaining in the container.

5. 75 mi $=$ _____ km

$\dfrac{75 \text{ mi}}{1}\left(\dfrac{1.6093 \text{ km}}{1 \text{ mi}}\right) = 120.6975$ km

7. 4 L $=$ _____ pt

$\dfrac{4 \text{ L}}{1}\left(\dfrac{1.0567 \text{ qt}}{1 \text{ L}}\right)\left(\dfrac{2 \text{ pt}}{1 \text{ qt}}\right) = 8.4536$ pt

9. $48°$F $=$ _____ $°$C

$°\text{C} = \dfrac{5}{9}\,(°\text{F} - 32)$

$°\text{C} = \dfrac{5}{9}\,(48 - 32)$

$°\text{C} = \dfrac{5}{9}\,(16)$

$°\text{C} = 8.888888889$

$°\text{C} = 9°\text{C}$ (rounded)

11. $42 \text{ h}\left(\dfrac{60 \text{ min}}{1 \text{ h}}\right) = 2{,}520$ min

13. $\dfrac{47\ \text{h}}{\text{week}}$ (5 weeks) = 235 h

15. 840 has 2 significant digits.

17. Absolute error = |24.75 cm − 24.72 cm|= 0.03 cm

Relative error = $\dfrac{0.03\ \text{cm}}{24.72\ \text{cm}}$ = 0.0012 (rounded)

Percent error = 0.0012 × 100% = 0.12%

19. 99 mm or 9.9 cm

21. $2{,}450\ \dfrac{\text{ft}}{\text{s}} = \underline{\quad}\ \dfrac{\text{m}}{\text{s}}$

$\dfrac{2{,}450\ \text{ft}}{\text{sec}}\left(\dfrac{0.3048\ \text{m}}{1\ \text{ft}}\right) = 746.76\ \dfrac{\text{m}}{\text{s}}$

The bullet travels $747\ \dfrac{\text{m}}{\text{s}}$.

23.
$$\begin{array}{r} 3.8\ \text{cm} \\ +5.9\ \text{cm} \\ \hline 9.7\ \text{cm} \end{array}$$

$9.7 \div 2 = 4.85$ cm

chapter 5

Signed Numbers and Powers of 10

Chapter Review Exercises

1. $|5| = 5$

3. $|+7| = 7$

5. The opposite of -12 is $+12$.

7. The opposite of -2 is $+2$.

9. The opposite of $+87$ is -87.

11. $(-15) + 8 = -7$
Different signs, subtract, keep the sign of the larger absolute value.

13. $-25 + 0 + 12 + 7 =$
$-25 + 12 + 7 =$
$-13 + 7 =$
-6

15. $+4 - 5 + 9 = -1 + 9 = 8$;
the team gained a net of $+8$ yards.

17. $\$500 - \$42 - \$18 - \$21 + \$150 =$
$458 - 18 - 21 + 150 =$
$440 - 21 + 150 =$
$419 + 150 =$
$\$569$

19. $-9 - 4 =$ Change subtraction sign to addition.
$-9 + (-4) =$ Change sign of subtrahend.
-13 Add.

21. $11 - (-3) =$ Change subtraction sign to addition.
$11 + (+3) =$ Change sign of subtrahend.
14 Add.

23. $\underline{-6 + 3} - 5 - 7 =$ Change subtraction sign to addition.
               Change sign of subtrahend.
$\underline{-3 - 5} - 7 =$ Add.
$\underline{-3 + (-5)} - 7 =$
$-8 - 7 =$
$-8 + (-7) =$
-15

25. $(+43) - (-27) =$ Difference in $43°$ above $0°$ $(+43)$ and $27°$ below $0°$ (-27)
$43 + (+27) =$ Change signs to addition.
$70°\text{F}$ Add.

27. $97.8° - 103.2° = -5.4°\text{F}$

29. $14° - (-22°) = 14° + 22° = 36°\text{C}$

31. $4,805 - 5,103 = -298$
298 million dollar decrease

33. $6,008 - 6,141 = -133$
133 million dollar decrease

35. $7(-2) = -14$
Different signs give negative product.

37. $2(3)(-7)(0) = 0$ Multiplying by zero results in zero.

39. $4(3)(-2)(7) = -168$ Since the number of negative factors is odd, the product will be negative.

41. $(7)^3 = (7)(7)(7) = 343$ A positive number raised to a power is positive.

43. $-4^2 = -(4 \cdot 4) = -16$ The exponent does not apply to the negative sign.

45. $5^2 = 5 \cdot 5 = 25$ A positive number raised to a power is positive.

47. $-2°(5) = -10°$ Product of drop in temperature ($-2°$) and hours observed (5).

49. Higher, 20 pts.   Drop of 4 points (-4) over a period of 5 weeks.
     $-4(5) = -20$

51. $12 \div 3 = 4$ **53.** $\dfrac{-20}{-5} = +4$
     Same signs give positive quotient.
                                         Same signs give a positive quotient.

55. $\dfrac{-51}{-3} = +17$ **57.** $\dfrac{-7}{0} =$ undefined
     Same signs give a positive quotient.
                                         Division of a nonzero integer by zero is undefined.

59. $\dfrac{51}{-17} = -3$ **61.** $7(3+5) =$ Parentheses or grouping.
     Different signs give a negative quotient. $7(8) =$ Multiply.
                                              56

                                         one calculator option:

63. $\dfrac{15-7}{8} =$ Parentheses or grouping. $7\ \boxed{(}\ 3\ \boxed{+}\ 5\ \boxed{)}\ \boxed{=}$

     $\dfrac{8}{8} =$ Division.   another calculator option:

     1 $7\ \boxed{\times}\ \boxed{(}\ 3\ \boxed{+}\ 5\ \boxed{)}\ \boxed{=}$

65.

$$4 + (-3)^4 - 2(5 + 1) =$$ Parentheses.
$$4 + (-3)^4 - 2(6) =$$ Exponentiation.
$$4 + 81 - 2(6) =$$ Multiply.
$$4 + 81 - 12 =$$ Add.
$$85 - 12 =$$ Subtract.
$$73$$

one calculator option:

4 | + | (| (−) | 3 |) | ∧ | 4 | − | 2 | (| 5 | + | 1 |) | ENTER

another calculator option:

4 | + | 3 | +/− | x^y | 4 | − | 2 | × | (| 5 | + | 1 |) | =

67. $\dfrac{-4}{5} \div \dfrac{-7}{15} = \dfrac{-4}{\cancel{5}_1} \times \dfrac{\cancel{15}^3}{-7} = \dfrac{-12}{-7} = \dfrac{12}{7}$ or $1\dfrac{5}{7}$

69. $-12.4 \div 0.2 = 0.2\overline{)-12.4} = -62$

$$\begin{array}{r} -62 \\ \hline -12.4 \\ 12 \\ \hline 04 \\ 4 \\ \hline 0 \end{array}$$

71. $-\dfrac{7}{8} + \left(-\dfrac{5}{12}\right) = \dfrac{-7}{8} + \left(\dfrac{-5}{12}\right) = \dfrac{-21}{24} + \left(\dfrac{-10}{24}\right) = \dfrac{-31}{24}$ or $-1\dfrac{7}{24}$

73. $-2\dfrac{5}{8} \times 4\dfrac{1}{2} = \dfrac{-21}{8} \times \dfrac{9}{2} = \dfrac{-189}{16}$ or $-11\dfrac{13}{16}$

75. $0.2 - 3.1(-7.6) = 0.2 + 23.56 = 23.76$

77. $10^5 \cdot 10^7 = 10^{5+7} = 10^{12} = 1{,}000{,}000{,}000{,}000$

79. $10^7 \cdot 10^{-10} = 10^{7+(-10)} = 10^{-3} = \dfrac{1}{1000} = 0.001$

81. $8.73 \div 10^{-3} = 8.73 \times 10^3$
$$= 8.730$$
$$= 8{,}730.$$
or
$$8.73 \div 10^{-3} = 8.73 \div \dfrac{1}{1{,}000}$$
$$= 8.73 \times \dfrac{1{,}000}{1}$$
$$= 8{,}730$$

83. $3.75 \times 10^5 =$
375,000

85. $3.87 \times 10^{-5} =$
0.0000387

87. $52{,}000 \to 5{,}2000$
5.2×10^4

89. $0.00017 \to 0.0001{_\wedge}7$
1.7×10^{-4}

91. $0.000\,000\,008 \to 0.000\,000\,008{_\wedge}$
8×10^{-9}

93. $\dfrac{1.25 \times 10^3}{3.7 \times 10^{-8}}$ $= 0.3378378378 \times 10^{11}$
$= 3.4 \times 10^{-1} \times 10^{11}$ (rounded)
$= 3.4 \times 10^{10}$

95. $(8.3 \times 10^{-2})^3$
$= 8.3^3 \times (10^{-2})^3$
$= 571.787 \times 10^{-6}$
$= 5.71787 \times 10^2 \cdot 10^{-6}$
$= 5.7 \times 10^{-4}$

97. $250{,}000{,}000 \rightarrow 2{,}50{,}000{,}000$
2.5×10^8

99. $0.00000092 = 920 \times 10^{-9}$
Exponent of 10 must be a multiple of 3.

101. $8{,}400{,}000 = 8.4 \times 10^6$

103. $41 = 41 \times 10^0$

105. $17{,}000{,}000 = 17 \times 10^6$

107. $3{,}084{,}000{,}000 = 3.084 \times 10^9$

109. $0.0000018 = 1.8 \times 10^{-6}$

111. $0.007 = 7 \times 10^{-3}$

113. $0.00000000035 = 350 \times 10^{-12}$

115. $0.00049 \text{ s} = 490 \times 10^{-6} \text{ s} = 490\,\mu\text{s}$

117. $0.588 \text{ Å} = 588 \times 10^{-3} \text{Å} = 588 \text{ mÅ}$

119. $246.7 \text{ V} = 246.7 \times 10^0 \text{V} = 246.7 \text{ V}$

121. $42{,}000 = 42 \times 10^3 \text{mW} = 42 \text{ W}$

123. $5{,}729\,\mu\text{W} = 5.729 \times 10^3 \mu\text{W}$
$= 5.729 \times 10^3 \times 10^{-6} \text{ W}$
$= 5.729 \times 10^{-3} \text{ W}$
$= 5.729 \text{ mW}$

125. $4{,}800 \text{ GHz} = 4.8 \times 10^3 \text{GHz}$
$= 4.8 \times 10^3 \times 10^9 \text{ Hz}$
$= 4.8 \times 10^{12} \text{ Hz}$
$= 4.8 \text{ THz}$

Chapter 5 Practice Test

1. $-8 < 0$ **3.** $-5 > -10$ **5.** The opposite of 8 is -8.

7. $-3 + 7 =$ Adding unlike signs: subtract,
      4   the sum has sign of number with larger absolute value.

9. $2(6)(-4) =$
　　　$12(-4) =$　　　Multiplying integers with unlike signs:
　　　　-48　　　multiply absolute values, negative product.

11. $-1.3 - 2.4 + 5.8 =$
　　　　$-3.7 + 5.8 =$
　　　　　　2.1

13. $(-8)(3)(0)(-1) = 0$
　　　Multiplying by zero results in zero.

15. $\dfrac{-7}{0}$　　　Undefined

17. $\dfrac{4}{-2} = -2$　　　Dividing numbers with unlike
　　　　　　　　signs gives negative quotient.

19. $2(3 - 9) \div 2^2 + 7 =$　　Parentheses and exponentiation.
　　　$2(-6) \div 4 + 7 =$　　Multiply.
　　　　$-12 \div 4 + 7 =$　　Divide.
　　　　　$-3 + 7 =$　　Add.
　　　　　　　4

21. $(10^3)^2 = 10^{3(2)} = 10^6$

23. $42 \times 10^3 = 42.\underset{\curvearrowright}{000} = 42{,}000$

25. $5.9 \times 10^{-2} = 0.059$

27. $0.021 = 0.02_{\wedge}1 = 2.1 \times 10^{-2}$

29. $783 \times 10^{-5} = 7_{\wedge}83 \times 10^{-5}$
　　　　　$= 7.83 \times 10^2 \times 10^{-5}$
　　　　　$= 7.83 \times 10^{-3}$

31. $\dfrac{5.25 \times 10^4}{1.5 \times 10^2} = \dfrac{5.25}{1.5} \cdot \dfrac{10^4}{10^2} = 3.5 \times 10^2$

33. $R = \dfrac{V}{A}$

　　　$R = \dfrac{3 \times 10^3 \ V}{2 \times 10^{-3} \ A}$

　　　$R = 1.5 \times 10^6 \ \Omega$

　　　$R = 1{,}500{,}000 \ \Omega$

35. $135°F - (-40°F) =$
　　　$135°F + 40°F =$
　　　　　$175°F$

37. $0.0047 \ G\Omega =$
　　　$4.7 \times 10^{-3} \ G\Omega =$
　　　$4.7 \times 10^{-3} \times 10^9 \ \Omega =$　　Exponent of
　　　$4.7 \times 10^6 \ \Omega =$　　　10 must be a
　　　$4.7 \ M\Omega$　　　　　　multiple of 3.

39. $0.000000023 \ fs =$
　　　$23 \times 10^{-9} \ fs =$　　Exponent of
　　　$23 \times 10^{-9} \times 10^{-15} \ s =$　　10 must be a
　　　$23 \times 10^{-24} \ s =$　　　multiple of 3.
　　　　$23 \ ys$

chapter 6   Statistics

Chapter Review Exercises

1. Women used more sick days than men in 2004, 2006, and 2007.

3. Men used more sick days than women in 2005, 2008, and 2009.

5. $R = \dfrac{8,500}{66,000}(100\%)$

$R = \dfrac{850,000}{66,000}\%$

$R = 12.9\%$

7.
$$\begin{array}{r} 8,500 \\ +\ 3,000 \\ \hline 11,500 \end{array}$$

$R = \dfrac{11,500}{66,000}(100\%)$

$R = \dfrac{1,150,000}{66,000}\%$

$R = 17.4\%$

9. 7-10-2009 at 4:00 PM

11. 3,560,000 + 1,800,000 + 480,000 + 550,000 + 185,000 = 6,575,000 barrels.

13. 1,800,000 ÷ 6,575,000 × 100% = 27.4 % (nearest tenth)

15. 1997; Locate the highest point on the graph.

17. 1995; Locate the steepest segment sloping upward to the right.
770,000 – 710,000 = 60,000 acres

19. 12,930 – 11,545 = 1,385
$\dfrac{1,385}{11,545}$ (100%) = 12% (nearest whole percent)

21. 2003; The graph shows thousand tons. So 11,545 on the graph indicates production of 11,545,000 tons.

23.
$$\begin{array}{r} 12.6 \\ 5\overline{\smash)63.0} \\ \underline{5} \\ 13 \\ \underline{10} \\ 30 \\ \underline{30} \\ \end{array}$$
The automobile salesperson sold an average of 12.6 cars per month.

25.
$$\begin{array}{r} 13.76 \approx 13.8 \\ 21\overline{\smash)289.00} \\ \underline{21} \\ 79 \\ \underline{63} \\ 160 \\ \underline{147} \\ 130 \\ \underline{126} \\ 4 \end{array}$$
The delivery truck used an average of 13.8 miles per gallon.

27. $11.90
$10.20 $\Big\}$ $\dfrac{10.20 + 9.85}{2} = \dfrac{20.05}{2}$
$9.85 $
$9.45   = $10.03 median

29.   The mode is $1.85, as it is the most frequent price.

31.

Points			
Hours	per	Hour	
3	×	4	= 12
3	×	3	= 9
+ 3	×	4	= + 12
9			33

$\begin{array}{r} 3.666 \approx 3.67 \\ 9\overline{)33.000} \\ \underline{27} \\ 60 \\ \underline{54} \\ 60 \\ \underline{54} \\ 60 \\ \underline{54} \\ 6 \end{array}$

The student's QPA is 3.67.

33.

$\begin{array}{r} 90 \\ \times\ \ 6 \\ \hline 540 \end{array}$ $\begin{array}{r} 99 \\ 93 \\ 91 \\ 88 \\ +\ 86 \\ \hline 457 \end{array}$ $\begin{array}{r} 540 \\ -\ 457 \\ \hline 83 \end{array}$

An 83 is needed on the last test for a 90 A average.

35.

Class Interval	Midpoint	Tally	Class Frequency
56-65	60.5	ЖΉ ЖΉ	10

37.

Class Interval	Midpoint	Tally	Class Frequency
36-45	40.5	ЖΉ ЖΉ II	12

39.

Class Interval	Midpoint	Tally	Class Frequency
16-25	20.5	ЖΉ II	7

41.   The age group 36-45 has the greatest number of members with 12 members.

43. $\begin{array}{r} 10 \\ +\ 5 \\ \hline 15 \end{array}$

There are 15 members over age 55.

45. $\dfrac{5}{10} = \dfrac{1}{2}$

The ratio of the number of members age 66 to 75 to members age 46 to 55 is $\dfrac{1}{2}$.

47.
$$R = \frac{10}{54}(100\%)$$

$$R = \frac{1,000}{54}\%$$

$$R = 18.5\%$$

49.

miles per gallon	Midpoint	Tally	MPG Frequency
20-24	22	ⅡⅡⅡ Ⅱ	7
25-29	27	ⅡⅡⅡ Ⅰ	6
30-34	32	ⅠⅠⅠⅠ	4

51. $\sum f = 7 + 6 + 4 = 17$

$\sum xf = 7(22) + 6(27) + 4(32) = 444$

mean (grouped) $= \dfrac{\sum xf}{\sum f} = \dfrac{444}{17} = 26.1$

53. Range $= 98 - 39 = 59$

55. $\text{Mean} = \dfrac{490}{6} = 81.66666667$

x	$\overline{x}$	$x - \overline{x}$	$(x - \overline{x})^2$
98	81.66666667	16.33333333	266.7777777
92	81.66666667	10.33333333	106.7777777
90	81.66666667	8.33333333	69.44444439
88	81.66666667	6.33333333	40.11111107
83	81.66666667	1.33333333	1.77777778
39	81.66666667	-42.66666667	1820.444445
490		0	2305.333334

$s = \sqrt{\dfrac{2305.333334}{6 - 1}} = \sqrt{461.0666667} = 21.47$

57.

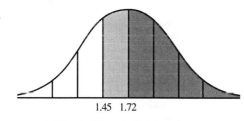

1.45 1.72

$1.72 - 0.27 = 1.45$

$50\% + 34.13\% = 84.13\%$

$1,000 \times 84.13\% =$

$1,000 \times 0.8413 =$

841.3

Approximately 841 men will be taller than 1.45 m.

59.

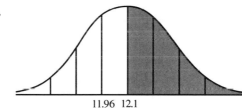

11.96 12.1

$12.1 - 11.9 = 0.2$

12.1 is one standard deviation above the mean.

$0.5 - 0.3413 = 0.1587$

15.87% of the containers are filled to 12.1 oz or more.

$3,000(15.87\%) =$

$3,000(0.1587) =$

476.1

476 containers are filled to 12.1 oz or more.

61. $\dfrac{10}{13}$ probability of choosing a fashion magazine on the first draw

$\dfrac{10 - 1}{13 - 1} = \dfrac{9}{12} = \dfrac{3}{4}$ probability of choosing a fashion magazine on the second draw if a fashion magazine was selected on the first draw.

63.

HHHH	THHH
HHHT	THHT
HHTH	THTH
HHTT	THTT
HTHH	TTHH
HTHT	TTHT
HTTH	TTTH
HTTT	TTTT

16 combinations

$2 \cdot 2 \cdot 2 \cdot 2 = 16$

Chapter 6 Practice Test

1. bar   **3.** line   **5.**
$$\begin{array}{r} 14 \\ -12 \\ \hline 2°\ C \end{array}$$
7.
$$\begin{array}{r} 50 \\ +\ 15 \\ \hline \$65 \end{array}$$
9.
$$R = \frac{50}{200}(100\%)$$
$$R = \frac{5,000}{200}\%$$
$$R = 25\%$$

11. $\dfrac{15}{200} = \dfrac{3}{40}$ **13.** Women make more than men in the English and Electronic departments. The bars for women in these two departments are longer than the bars for men.

15. $R = \dfrac{50}{48}(100\%)$ **17.** $\dfrac{2}{4} = \dfrac{1}{2}$ **19.** range: $93 - 68 = 25$

$$R = \frac{5,000}{48}\%$$
$$R = 104.1666667\%$$
$$R = 104\% \ (\text{rounded})$$

$$\begin{array}{r} 93 \\ 81 \\ 81 \\ 78 \\ 75 \\ 69 \\ +\ 68 \\ \hline 545 \end{array}$$
$\left.\begin{array}{l} \\ \\ \end{array}\right\} \leftarrow$ mode
$\leftarrow$ median

$$\begin{array}{r} 77.85 \approx 77.9\ \text{mean} \\ 7\,\overline{|\ 545.00} \\ \underline{49} \\ 55 \\ \underline{49} \\ 60 \\ \underline{56} \\ 40 \\ \underline{35} \\ 5 \end{array}$$

21. Mean $= \dfrac{81 + 78 + 69 + 75 + 81 + 93 + 68}{7} = \dfrac{545}{7} = 77.85714286$

x	$\overline{x}$	$x - \overline{x}$	$(x - \overline{x})^2$
81	77.85714286	3.142857143	9.87755102
78	77.85714286	0.14285714	0.0204081624
69	77.85714286	− 8.85714286	78.44897964
75	77.85714286	− 2.85714286	8.163265322
81	77.85714286	3.14285714	9.877551002
93	77.85714286	15.14285714	229.3061224
68	77.85714286	− 9.85714286	97.16326536

$$\sum (x - \overline{x})^2 = 432.8571429$$

Standard deviation $= \sqrt{\dfrac{\sum (x - \overline{x})^2}{n - 1}}$

$$= \sqrt{\dfrac{432.8571429}{6}}$$
$$= \sqrt{72.14285715}$$
$$= 8.493695141$$
$$sd = 8.49 \ (\text{rounded})$$

23. $2 \times 3 = 6$

25.
$$\begin{array}{r} 2 \\ +\ 3 \\ \hline 5 \end{array}$$
$\dfrac{3}{5}$ probability of drawing a woman's name first.

27.

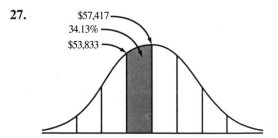

$\$57,417 - \$53,833 = \$3,584$
34.13% of hospital stays cost between $\$53,833$ and $\$57,417$ or one standard deviation below the mean.

Chapters 4 – 6 Cumulative Practice Test

1. 14.3 dm = 1.43 m

3. 30,003 dg = 3.002 kg

5. $86°F = $ _____ $°C$

$$°C = \frac{5}{9} (°F - 32)$$

$$°C = \frac{5}{9} (86 - 32)$$

$$°C = \frac{5}{9} (54)$$

$$°C = 5(6)$$

$$°C = 30°$$

7. The greatest possible error of $3\frac{1}{4}$ cm is $\frac{1}{2} \times \frac{1}{4}$ cm $= \frac{1}{8}$ cm.

9. 0.00356 has three significant digits. Zeros to the left of the first nonzero digit are not significant digits.

11.   5.7 cm
3.05 cm
21.46 cm
_____
30.21 cm rounds to 30.2 cm

13.
$$-3 - 8 - 5 - (-2) =$$
$$-3 + (-8) + (-5) + (2) =$$
$$-11 + (-5) + 2 =$$
$$-16 + 2 =$$
$$-14$$

15. $-24 \div 4 = -6$

17. $-5\frac{1}{2} + 4\frac{1}{4} = -5\frac{2}{4} + 4\frac{1}{4} = -1\frac{1}{4}$

19. $5.3 \times 10^{-2} = 5.3 \times 0.01 = 0.053$

21. $0.003_{\wedge}5 = 3.5 \times 10^{-3}$

23. $0.000145_{\wedge}7 = 145.7 \times 10^{-6}$

25. $\dfrac{4.6 \times 10^4}{2.3 \times 10^{-2}} = 2 \times 10^6$

27. Arkansas is the leftmost bar. Read the lower shaded portion of the bar. 64,000 cwts

29. The table at the bottom of the graph shows that Mississippi and Missouri have blanks in the blocks for medium grain.

31. China has the largest share of the circle.

33.   Mean $= \dfrac{583 + 417 + 456 + 498 + 523 + 456}{6} = 489$ (rounded)

Median $= \dfrac{456 + 498}{2} = 477$ (average of two middle scores)

Mode $= 456$ (score with greatest frequency)

35. $p = \dfrac{\text{number of ways event can occur}}{\text{total number of occurrences}} = \dfrac{8}{30} = \dfrac{4}{15}$

Chapter Review Exercises

1. $5 + 7 = 18 - 6$
$12 = 12$

3. $8 - 9 + 3^2 = -2^2 + 12$
$-1 + 9 = -4 + 12$
$8 = 8$

5. $x = 5 + 9$
$x = 14$

7. $y = \dfrac{48}{-6}$
$y = -8$

9. $\boxed{15x} - \boxed{\dfrac{3a}{7}} + \boxed{\dfrac{x-7}{5}}$

11. Five more than a number is two, or a number increased by 5 equals 2. Answers will vary.

13. A number divided by 8 is 7. Answers will vary.

15. $2x + 7 = 11$

17. $2(x + 8) = 40$

19. $3a + 2a = 5a$

21. $3(2y - 4) + y$
$6y - 12 + y$
$7y - 12$

23. $-3(u + 2) - 5$
$-3a - 6 - 5$
$-3a - 11$

25. $11 - 2(x - 5) =$
$11 - 2x + 10 =$
$21 - 2x$

27. $3 - (2x - 4y - 8) =$
$3 - 2x + 4y + 8 =$
$11 - 2x + 4y$

29. $15a + 3(5a - 7b + 4) =$
$15a + 15a - 21b + 12 =$
$30a - 21b + 12$

31. $x - 5 + 5 = 8 + 5$
$x + 0 = 13$
$x = 13$

33. $x - 8 + 8 = -10 + 8$
$x + 0 = -2$
$x = -2$

35. $x - 5 + 5 = 14 + 5$
$x + 0 = 19$
$x = 19$

37. $x + 7 = 10$
$x + 7 - 7 = 10 - 7$
$x + 0 = 3$
$x = 3$

39. $x - 3 = -4$
$x - 3 + 3 = -4 + 3$
$x + 0 = -1$
$x = -1$

41. $1 = a - 4$
$1 + 4 = a - 4 + 4$
$5 = a + 0$
$5 = a$
$a = 5$

43. $3x + 4 = 19$
$3x + 4 - 4 = 19 - 4$
$3x = 15$
$\dfrac{3x}{3} = \dfrac{15}{3}$
$x = 5$

45. $15 - 3x = -6$
$15 - 15 - 3x = -6 - 15$
$-3x = -21$
$\dfrac{-3x}{-3} = \dfrac{-21}{-3}$
$x = 7$

47. $-7 = 6x - 31$
$-7 + 31 = 6x - 31 + 31$
$24 = 6x$
$\dfrac{24}{6} = \dfrac{6x}{6}$
$4 = x$
$x = 4$

49. $5x - 12 = 9x$
$5x - 5x - 12 = 9x - 5x$
$-12 = 4x$
$\dfrac{-12}{4} = \dfrac{4x}{4}$
$-3 = x$
$x = -3$

51. $3x = 21$
$\dfrac{\cancel{3}x}{\cancel{3}_1} = \dfrac{\cancel{21}^7}{\cancel{3}}$
$x = 7$

53. $-15 = 2b$

$\dfrac{-15}{2} = \dfrac{2b}{2}$

$\dfrac{-15}{2} = b$

$b = \dfrac{-15}{2}$ or $-7\dfrac{1}{2}$ or -7.5

55. $3 = \dfrac{1}{5}x$

$\left(\dfrac{5}{1}\right)3 = \left(\dfrac{5}{1}\right)\dfrac{1}{5}x$

$15 = x$

$x = 15$

57. $-7y = -49$

$\dfrac{-7y}{-7} = \dfrac{-49}{-7}$

$y = 7$

Check:

$-7y = -49$

$-7(7) = -49$

$-49 = -49$

59. $-\dfrac{3}{8}x = -24$

$\left(\dfrac{-8}{3}\right)\dfrac{-3}{8}x = \left(\dfrac{-8}{3}\right)\dfrac{-24}{1}$

$x = 64$

Check:

$-\dfrac{3}{8}x = -24$

$-\dfrac{3}{8}(64) = -24$

$-24 = -24$

61. $42 = -\dfrac{6}{7}x$

$\left(\dfrac{-7}{6}\right)\left(\dfrac{42}{1}\right) = \left(\dfrac{-7}{6}\right)\left(\dfrac{-6}{7}x\right)$

$-49 = x$

$x = -49$

Check:

$42 = -\dfrac{6}{7}x$

$42 = \dfrac{-6}{7}(-49)$

$42 = 42$

63. $\dfrac{1}{7}x = 12$

$\left(\dfrac{7}{1}\right)\dfrac{1}{7}x = \left(\dfrac{7}{1}\right)\dfrac{12}{1}$

$x = 84$

Check:

$\dfrac{1}{7}x = 12$

$\dfrac{1}{7}(84) = 12$

$12 = 12$

65. $2b - 7b = 10$

$-5b = 10$

$\dfrac{-5b}{-5} = \dfrac{10}{-5}$

$b = -2$

67. $21 = x + 2x$

$21 = 3x$

$\dfrac{21}{3} = \dfrac{3x}{3}$

$7 = x$

$x = 7$

69. $4x + x = 25$

$5x = 25$

$\dfrac{5x}{5} = \dfrac{25}{5}$

$x = 5$

71. $20 - 4 = 2x - 6x$

$16 = -4x$

$\dfrac{16}{-4} = \dfrac{-4x}{-4}$

$-4 = x$

$x = -4$

73. $-12 = -8 - 2x$

$-12 + 8 = -8 + 8 - 2x$

$-4 = -2x$

$\dfrac{-4}{-2} = \dfrac{-2x}{-2}$

$2 = x$

$x = 2$

75. $3x + 9 = 10 + 3x + 1$

$3x + 9 = 11 + 3x$

$3x + 9 - 9 = 11 - 9 + 3x$

$3x = 2 + 3x$

$3x - 3x = 2 + 3x - 3x$

$0 = 2$ False

No real solutions

77. $7 - 4y = y + 22$

$-4y - y = 22 - 7$

$-5y = 15$

$\dfrac{-5y}{-5} = \dfrac{15}{-5}$

$y = -3$

79. $7x - 5 + 2x = 3 - 4x + 12$

$9x - 5 = 15 - 4x$

$9x + 4x = 15 + 5$

$13x = 20$

$\dfrac{13x}{13} = \dfrac{20}{13}$

$x = \dfrac{20}{13}$

81. $4y + 8 = 3y - 4$

$4y - 3y = -4 - 8$

$y = -12$

83. $8 - 2y = 15 - 3y$

$-2y + 3y = 15 - 8$

$y = 7$

85. $5x - 12 = 2x + 15$
$5x - 2x = 15 + 12$
$3x = 27$
$\dfrac{3x}{3} = \dfrac{27}{3}$
$x = 9$

87. $3x - 5x + 2 = 6x - 5 + 12x$
$-2x + 2 = 18x - 5$
$-2x - 18x = -5 - 2$
$-20x = -7$
$\dfrac{-20x}{-20} = \dfrac{-7}{-20}$
$x = \dfrac{7}{20}$ or 0.35

89. $0 = \dfrac{8}{9}c + \dfrac{1}{4}$
$0 - \dfrac{1}{4} = \dfrac{8}{9}c$
$-\dfrac{1}{4} = \dfrac{8}{9}c$
$\left(\dfrac{9}{8}\right)\left(-\dfrac{1}{4}\right) = \left(\dfrac{9}{8}\right)\dfrac{8}{9}c$
$c = -\dfrac{9}{32}$

91. $0.86 = R + 0.4R$
$0.86 = 1.4R$
$\dfrac{0.86}{1.4} = \dfrac{1.4R}{1.4}$
$0.61 = R$
$R = 0.61$

$\begin{array}{r} 1. \\ +\,0.4 \\ \hline 1.4 \end{array}$

$\begin{array}{r} 0.614 \quad \text{(round to hundredths)} \\ 1.4\,\overline{)0.860} \\ \underline{84} \\ 20 \\ \underline{14} \\ 60 \\ \underline{56} \\ 4 \end{array}$

93. $18 = 6(2 - y)$
$18 = 12 - 6y$
$18 - 12 = -6y$
$6 = -6y$
$\dfrac{6}{-6} = \dfrac{-6y}{-6}$
$-1 = y$
$y = -1$

95. $7x - 3(x - 8) = 28$
$7x - 3x + 24 = 28$
$4x = 28 - 24$
$4x = 4$
$\dfrac{4x}{4} = \dfrac{4}{4}$
$x = 1$

97. $5(3 - 2x) = -5$
$15 - 10x = -5$
$-10x = -5 - 15$
$-10x = -20$
$\dfrac{-10x}{-10} = \dfrac{-20}{-10}$
$x = 2$

99. $3x = 3(9 + 2x)$
$3x = 27 + 6x$
$3x - 6x = 27$
$-3x = 27$
$\dfrac{-3x}{-3} = \dfrac{27}{-3}$
$x = -9$

101. $5x = 7 + (x + 5)$
$5x = 7 + 1(x + 5)$
$5x = 7 + 1x + 5$
$5x - 1x = 7 + 5$
$4x = 12$
$\dfrac{4x}{4} = \dfrac{12}{4}$
$x = 3$

103. $4(3 - x) = 2x$
$12 - 4x = 2x$
$12 = 2x + 4x$
$12 = 6x$
$\dfrac{12}{6} = \dfrac{6x}{6}$
$2 = x$
$x = 2$

105. $-2(4 - 2x) = -16 + 2x$
$-8 + 4x = -16 + 2x$
$4x - 2x = -16 + 8$
$2x = -8$
$\dfrac{2x}{2} = \dfrac{-8}{2}$
$x = -4$

107.
$$8 = 6 - 2(3x - 1)$$
$$8 = 6 - 6x + 2$$
$$8 = 8 - 6x$$
$$8 - 8 = -6x$$
$$0 = -6x$$
$$\frac{0}{-6} = \frac{-6x}{-6}$$
$$0 = x$$
$$x = 0$$

109.
$$3(x - 1) = 18 - 2(x + 3)$$
$$3x - 3 = 18 - 2x - 6$$
$$3x - 3 = 12 - 2x$$
$$3x + 2x = 12 + 3$$
$$5x = 15$$
$$\frac{5x}{5} = \frac{15}{5}$$
$$x = 3$$

111.
$$-(2x + 1) = -7$$
$$-1(2x + 1) = -7$$
$$-2x - 1 = -7$$
$$-2x = -7 + 1$$
$$-2x = -6$$
$$\frac{-2x}{-2} = \frac{-6}{-2}$$
$$x = 3$$

113.
$$7 = 3 + 4(x + 2)$$
$$7 = 3 + 4x + 8$$
$$7 = 11 + 4x$$
$$7 - 11 = 4x$$
$$-4 = 4x$$
$$\frac{-4}{4} = \frac{4x}{4}$$
$$-1 = x$$
$$x = -1$$

115.
$$3(4x + 3) = 3 - 4(x - 1)$$
$$12x + 9 = 3 - 4x + 4$$
$$12x + 9 = 7 - 4x$$
$$12x + 4x = 7 - 9$$
$$16x = -2$$
$$\frac{16x}{16} = \frac{-2}{16}$$
$$x = -\frac{1}{8} \text{ or } -0.125$$

117.
$$x - 6 = 8$$
$$x = 8 + 6$$
$$x = 14$$

119.
$$5(x + 6) = x + 42$$
$$5x + 30 = x + 42$$
$$5x - x = 42 - 30$$
$$4x = 12$$
$$\frac{4x}{4} = \frac{12}{4}$$
$$x = 3$$

121.
$$x + (x - 3) = 51$$
$$x + x - 3 = 51$$
$$2x - 3 = 51$$
$$2x = 51 + 3$$
$$2x = 54$$
$$\frac{2x}{2} = \frac{54}{2}$$
$$x = 27$$
$$x - 3 = 27 - 3 = 24$$

One technician works 27 hours
and the other works 24 hours.

123.
$$P = 2(l + w); \quad l = 2w$$
$$720 = 2(2w + w)$$
$$720 = 4w + 2w$$
$$720 = 6w$$
$$\frac{720}{6} = \frac{6w}{6}$$
$$120 \text{ ft} = w$$
$$l = 2w = 2(120) = 240 \text{ ft}$$

125.
$$m + \frac{1}{4} = \frac{3}{4}$$
$$4m + 4\left(\frac{1}{4}\right) = 4\left(\frac{3}{4}\right)$$
$$4m + 1 = 3$$
$$4m = 3 - 1$$
$$4m = 2$$
$$m = \frac{2}{4}$$
$$m = \frac{1}{2}$$

127.
$$p = \frac{1}{2} + \frac{1}{3}$$
$$(3)(2)p = (3)(2)\frac{1}{2} + (3)(2)\frac{1}{3}$$
$$6p = 3 + 2$$
$$6p = 5$$
$$\frac{6p}{6} = \frac{5}{6}$$
$$p = \frac{5}{6}$$

129.
$$\frac{2}{5} - x = \frac{1}{2}x + \frac{4}{5}$$
$$(5)(2)\frac{2}{5} - (5)(2)x = (5)(2)\frac{1}{2}x + (5)(2)\frac{4}{5}$$
$$4 - 10x = 5x + 8$$
$$-10x - 5x = 8 - 4$$
$$-15x = 4$$
$$\frac{-15x}{-15} = \frac{4}{-15}$$
$$x = -\frac{4}{15}$$

131.
$$\frac{3}{7} m - \frac{1}{2} = \frac{2}{3}$$
$$(\cancel{7})(2)(3) \frac{3}{\cancel{7}} m - (7)(\cancel{2})(3) \frac{1}{\cancel{2}} = (7)(2)(\cancel{3}) \frac{2}{\cancel{3}}$$
$$18m - 21 = 28$$
$$18m = 28 + 21$$
$$18m = 49$$
$$\frac{18m}{18} = \frac{49}{18}$$
$$m = \frac{49}{18}$$

133.
$$m = 2 + \frac{1}{4} m$$
$$(4)m = (4)2 + (\cancel{4}) \frac{1}{\cancel{4}} m$$
$$4m = 8 + m$$
$$4m - m = 8$$
$$3m = 8$$
$$\frac{3m}{3} = \frac{8}{3}$$
$$m = \frac{8}{3}$$

135. Excluded value: $P \neq 0$
$$\frac{2}{P} = \frac{1}{2} + \frac{1}{4} - \frac{5}{12}$$
$$(12)(\cancel{P}) \frac{2}{\cancel{P}} = (1\cancel{2})(P) \frac{1}{\cancel{2}}^{6} + (\cancel{1}\cancel{2})(P) \frac{1}{\cancel{4}}^{3} - (\cancel{1}\cancel{2})(P) \frac{5}{\cancel{1}\cancel{2}}$$
$$24 = 6P + 3P - 5P$$
$$24 = 4P$$
$$\frac{24}{4} = \frac{4P}{4}$$
$$P = 6$$

137.
$$\frac{1}{3} x + \frac{1}{7} x = 1$$
$$(\cancel{3})(7) \frac{1}{\cancel{3}} x + (3)(\cancel{7}) \frac{1}{\cancel{7}} x = (3)(7)1$$
$$7x + 3x = 21$$
$$10x = 21$$
$$\frac{10x}{10} = \frac{21}{10}$$
$$x = \frac{21}{10} \text{ or } 2\frac{1}{10} \text{ h}$$
$$\text{or } 2.1 \text{ h}$$
$$\text{or } 2 \text{ h } 6 \text{ min}$$

139. 5 fixtures in 2 h $= \frac{5}{2}$ fixtures per hour
$$\left(\frac{5 \text{ fixtures}}{2 \text{ h}} \right)(10 \text{ h}) = 25 \text{ fixtures}$$

141.
$$\frac{1}{R} = \frac{1}{2} + \frac{1}{6} + \frac{1}{12}$$
$$(12R) \frac{1}{R} = (12R)\left(\frac{1}{2}\right) + (12R)\left(\frac{1}{6}\right) + (12R)\left(\frac{1}{12}\right)$$
$$12 = 6R + 2R + R$$
$$\frac{12}{9} = \frac{9R}{9}$$
$$\frac{12}{9} = R$$
$$R = \frac{4}{3}$$
$$R = 1\frac{1}{3} \text{ ohms or } 1.33 \text{ ohms}$$

143.
$$2.3x - 4.1 = 0.5$$
$$(10)2.3x - (10)4.1 = (10)0.5$$
$$23x - 41 = 5$$
$$23x = 5 + 41$$
$$23x = 46$$
$$\frac{23x}{23} = \frac{46}{23}$$
$$x = 2$$

145.
$$0.3x - 2.15 = 0.8x + 3.75$$
$$(100)0.3x - (100)2.15 = (100)0.8x + (100)3.75$$
$$30x - 215 = 80x + 375$$
$$30x - 80x = 375 + 215$$
$$-50x = 590$$
$$\frac{-50x}{-50} = \frac{590}{-50}$$
$$x = \frac{590}{-50}$$
$$x = -11.8$$

147.
$$R = \frac{V}{A}$$
$$R = \frac{8.5}{0.5}$$
$$R = 17$$

149. $I = \$387.50,\ P = \$1,550,\ R = 12.5\%$
$$I = PRT$$
$$387.50 = 1,550(0.125)T$$
$$387.50 = 193.75T$$
$$\frac{387.50}{193.75} = \frac{193.75T}{193.75}$$
$$2 = T$$
$$T = 2 \text{ yr}$$

151.
$$E = IR$$
$$220 = I(80)$$
$$\frac{220}{80} = \frac{I(80)}{80}$$
$$2.75 = I$$
$$I = 2.75 \text{ A}$$

153.
$$P = I^2R$$
$$392 = I^2(8)$$
$$\frac{392}{8} = \frac{I^2(8)}{8}$$
$$49 = I^2$$
$$\sqrt{49} = I$$
$$I = 7 \text{ amps}$$

155. $V = lwh \quad$ for w
$$\frac{V}{lh} = \frac{lwh}{lh}$$
$$\frac{V}{lh} = w$$
$$w = \frac{V}{lh}$$

157. $s = r - d \quad$ for r
$$s + d = r$$
$$r = s + d$$

159. $v = v_0 - 32t \quad$ for t
$$v - v_0 = -32t$$
$$\frac{v - v_0}{-32} = \frac{-32t}{-32}$$
$$\frac{v - v_0}{-32} = t \text{ or } \frac{v_0 - v}{32} = t$$
$$t = \frac{v_0 - v}{32}$$

161. $S = P - D \quad$ for P
$$S + D = P$$
$$P = S + D$$

Chapter 7 Practice Test

1.
$$x + 5 - 5 = 19 - 5$$
$$x = 19 - 5$$
$$x = 14$$

3.
$$\frac{x}{2} = 5$$
$$\left(\frac{2}{1}\right)\left(\frac{x}{2}\right) = \left(\frac{2}{1}\right)\left(\frac{5}{1}\right)$$
$$x = 10$$

5.
$$5 - 2x = 3x - 10$$
$$-2x - 3x = -10 - 5$$
$$-5x = -15$$
$$\frac{-5x}{-5} = \frac{-15}{-5}$$
$$x = 3$$

7.
$$3(x + 4) = 18$$
$$3x + 12 = 18$$
$$3x = 18 - 12$$
$$3x = 6$$
$$\frac{3x}{3} = \frac{6}{3}$$
$$x = 2$$

9.
$$\frac{8}{y+2} = -7 \qquad \text{Excluded value: } y + 2 = 0$$
$$y = -2$$
$$(y+2)\frac{8}{y+2} = \frac{-7}{1}(y+2)$$
$$8 = -7(y+2)$$
$$8 = -7y - 14$$
$$8 + 14 = -7y$$
$$22 = -7y$$
$$\frac{22}{-7} = \frac{-7y}{-7}$$
$$y = -\frac{22}{7}$$

11.
$$5x + \frac{3}{5} = 2$$
$$(5)5x + (\cancel{5})\frac{3}{\cancel{5}} = (5)2$$
$$25x + 3 = 10$$
$$25x = 10 - 3$$
$$25x = 7$$
$$\frac{25x}{25} = \frac{7}{25}$$
$$x = \frac{7}{25} \text{ or } 0.28$$

13.
$$\frac{3}{5}x + \frac{1}{10}x = \frac{1}{3}$$
$$(\cancel{30})^6 \frac{3}{\cancel{5}} x + (\cancel{30})^3 \frac{1}{\cancel{10}} x = (\cancel{30})^{10} \frac{1}{\cancel{3}}$$
$$18x + 3x = 10$$
$$21x = 10$$
$$\frac{21x}{21} = \frac{10}{21}$$
$$x = \frac{10}{21}$$

15.
$$1.3x = 8.02$$
$$(100)1.3x = (100)8.02$$
$$130x = 802$$
$$x = 6.169230769$$
$$x = 6.17$$

17.
$$0.18x = 300 - x$$
$$100(0.18x) = 100(300) - 100x$$
$$18x = 30,000 - 100x$$
$$18x + 100x = 30,000$$
$$118x = 30,000$$
$$x = 254.2372881$$
$$x = 254.24$$

19.
$$0.23 + 7.1x = -0.8$$
$$100(0.23) + 100(7.1x) = 100(-0.8)$$
$$23 + 710x = -80$$
$$710x = -80 - 23$$
$$710x = -103$$
$$\frac{710x}{710} = \frac{-103}{710}$$
$$x = -0.1450704225$$
$$x = -0.15$$

21.
$$E = IR$$
$$110 \text{ V} = I(50 \ \Omega)$$
$$\frac{110}{50} = \frac{I(50)}{50}$$
$$2.2 \text{ A} = I$$
$$\text{or}$$
$$I = 2.2 \text{ A}$$

23.
$$R = \frac{PL}{A}$$
$$(A)R = (\cancel{A})\frac{PL}{\cancel{A}}$$
$$AR = PL$$
$$\frac{AR}{P} = \frac{PL}{\cancel{P}}$$
$$\frac{AR}{P} = L$$
$$L = \frac{AR}{P}$$

25.
$$d = \pi r^2 sn$$
$$\frac{d}{\pi r^2 n} = \frac{\pi r^2 sn}{\pi r^2 n}$$
$$\frac{d}{\pi r^2 n} = s$$
$$s = \frac{d}{\pi r^2 n}$$

27.
$$E = \frac{I - P}{I}$$
$$0.7 = \frac{40,000 - P}{40,000}$$
$$(40,000)\frac{0.7}{1} = \frac{40,000 - P}{40,000}(40,000)$$
$$28,000 = 40,000 - P$$
$$28,000 - 40,000 = -P$$
$$-12,000 = -P$$
$$\frac{-12,000}{-1} = \frac{-P}{-1}$$
$$12,000 = P$$
$$P = 12,000 \text{ calories}$$

Chapter Review Exercises

1.
$$\frac{7}{x} = 6$$
$$\frac{7}{x} = \frac{6}{1}$$
$$7 = 6x$$
$$\frac{7}{6} = \frac{6x}{6}$$
$$x = \frac{7}{6}$$

3.
$$\frac{4x+3}{15} = \frac{1}{3}$$
$$12x + 9 = 15$$
$$12x = 15 - 9$$
$$12x = 6$$
$$\frac{12x}{12} = \frac{6}{12}$$
$$x = \frac{1}{2}$$

5.
$$\frac{5}{4x-3} = \frac{3}{8}$$
$$5(8) = 3(4x - 3)$$
$$40 = 12x - 9$$
$$40 + 9 = 12x$$
$$49 = 12x$$
$$\frac{49}{12} = \frac{12x}{12}$$
$$\frac{49}{12} = x$$
$$x = \frac{49}{12}$$

7.
$$\frac{4x}{7} = \frac{2x+3}{3}$$
$$4x(3) = 7(2x + 3)$$
$$12x = 14x + 21$$
$$12x - 14x = 21$$
$$-2x = 21$$
$$\frac{-2x}{-2} = \frac{21}{-2}$$
$$x = -\frac{21}{2}$$

9.
$$\frac{5x}{3} = \frac{2x+1}{4}$$
$$20x = 6x + 3$$
$$20x - 6x = 3$$
$$14x - 3$$
$$\frac{14x}{14} = \frac{3}{14}$$
$$x = \frac{3}{14}$$

11.
$$\frac{7}{x} = \frac{5}{4x+3}$$
$$7(4x + 3) = 5x$$
$$28x + 21 = 5x$$
$$21 = 5x - 28x$$
$$21 = -23x$$
$$\frac{21}{-23} = \frac{-23x}{-23}$$
$$x = -\frac{21}{23}$$

13.
$$\frac{25}{x} = \frac{35}{6,300}$$
$$35x = 25(6,300)$$
$$35x = 157,500$$
$$35x = \frac{157,500}{35}$$
$$x = 4,500 \text{ women}$$

15.
$$\frac{\frac{5}{8}}{2} = \frac{1\frac{5}{16}}{x}$$
$$\frac{\frac{5}{8}}{2} = \frac{\frac{21}{16}}{x}$$
$$\frac{5}{8}x = \overset{1}{\cancel{2}}\left(\frac{21}{\underset{8}{\cancel{16}}}\right)$$
$$\frac{5}{8}x = \frac{21}{8}$$
$$\left(\frac{\overset{1}{\cancel{8}}}{\cancel{5}}\right)\frac{\overset{1}{\cancel{5}}}{\cancel{8}}x = \left(\frac{\overset{1}{\cancel{8}}}{5}\right)\frac{21}{\underset{1}{\cancel{8}}}$$
$$x = \frac{21}{5}$$
$$x = 4\frac{1}{5} \text{ ft}$$

17.
$$\frac{81.2}{x} = \frac{845}{1,350}$$
$$845x = 81.2(1,350)$$
$$845x = 109,620$$
$$\frac{845x}{845} = \frac{109,620}{845}$$
$$x = 129.7278107$$
$$x = 129.7 \text{ gal}$$

19.
$$y = kx$$
$$3.5 = k(300)$$
$$k = \frac{3.5}{300}$$
$$k = 0.011666667$$
$$y = 0.011666667x$$
$$y = 0.011666667(425)$$
$$y = 5.0h$$

21.
$$y = kx$$
$$825 = k(1.983)$$
$$k = \frac{825}{1.983}$$
$$k = 416.0363086$$
$$y = 416.0363086x$$
$$y = 416.0363086(3.247)$$
$$y = 1,351 \text{ ft}$$

23. $b = kde$
$b = 0.8(12)(25)$
$b = 240$

25. $A = kbh$
$225 = k(18)(25)$
$225 = 450k$
$k = \dfrac{225}{450}$
$k = 0.5$
$A = 0.5bh$
$A = 0.5(35)(40)$
$A = 700 \text{ in}^2$

27. $\dfrac{15}{x} = \dfrac{4}{6}$
$4x = 90$
$\dfrac{4x}{4} = \dfrac{90}{4}$
$x = \dfrac{90}{4}$
$x = \dfrac{45}{2} \text{ or } 22\dfrac{1}{2}$
23 machines

29. $\dfrac{50}{40} = \dfrac{x}{2}$
$40x = 100$
$\dfrac{40x}{40} = \dfrac{100}{40}$
$x = \dfrac{100}{40}$
$x = 2\dfrac{1}{2} \text{ h}$

31. $\dfrac{45}{30} = \dfrac{x}{1000}$
$30x = 45,000$
$\dfrac{30x}{30} = \dfrac{45,000}{30}$
$x = 1,500 \text{ rpm}$

33. $y = \dfrac{k}{x}$
$5 = \dfrac{k}{3}$
$k = 5(3)$
$k = 15$
$y = \dfrac{15}{5}$
$y = 3 \text{ da}$

35. $y = \dfrac{k}{x}$
$30 = \dfrac{k}{4}$
$k = 30(4)$
$k = 120$
$y = \dfrac{120}{x}$
$y = \dfrac{120}{2}$
$y = 60 \text{ rpm}$

37. $P = \dfrac{kr}{s^2}$ $P = \dfrac{25r}{s^2}$
$175 = \dfrac{k(252)}{6^2}$ $P = \dfrac{25(196)}{7^2}$
$175 = \dfrac{252k}{36}$ $P = \dfrac{4,900}{49}$
$36(175) = 252k$ $P = 100$
$6,300 = 252k$
$k = \dfrac{6,300}{252}$
$k = 25$

Chapter 8 Practice Test

1. $\dfrac{x}{12} = \dfrac{5}{8}$
$8x = 5(12)$
$8x = 60$
$\dfrac{8x}{8} = \dfrac{60}{8}$
$x = \dfrac{15}{2}$

3. $\dfrac{640}{24} = \dfrac{x}{360}$
$24x = 640(360)$
$24x = 230,400$
$\dfrac{24x}{24} = \dfrac{230,400}{24}$
$x = 9,600$

5. $\dfrac{2\dfrac{1}{2}}{x} = \dfrac{1\dfrac{1}{4}}{3\dfrac{1}{5}}$
$1\dfrac{1}{4}x = 2\dfrac{1}{2}\left(3\dfrac{1}{5}\right)$
$\dfrac{5}{4}x = \dfrac{5}{2}\left(\dfrac{16}{5}\right)$
$\dfrac{5}{4}x = 8$
$\left(\dfrac{4}{5}\right)\dfrac{5}{4}x = 8\left(\dfrac{4}{5}\right)$
$x = \dfrac{32}{5}$

7. Let x = rpm of smaller gear
$\dfrac{300 \text{ teeth}}{60 \text{ teeth}} = \dfrac{x \text{ rpm}}{40 \text{ rpm}}$
$60x = 300(40)$
$60x = 12,000$
$\dfrac{60x}{60} = \dfrac{12,000}{60}$
$x = 200 \text{ rpm}$

9. inverse proportion

$$\frac{9}{4} = \frac{x}{75}$$

$$4x = 675$$

$$\frac{4x}{4} = \frac{675}{4}$$

$$x = \frac{675}{4} \text{ or } 168\frac{3}{4} \text{ or } 168.75 \text{ rpm}$$

11. direct proportion

$$\frac{32.75}{117.9} = \frac{24.65}{x}$$

$$32.75x = 117.9(24.65)$$

$$32.75x = 2,906.235$$

$$x = \frac{2,906.235}{32.75}$$

$$x = 88.74 \text{ in}^2$$

13. direct proportion

$$\frac{62.5}{x} = \frac{400}{350}$$

$$400x = 62.5(350)$$

$$400x = 21,875$$

$$\frac{400x}{400} = \frac{21,875}{400}$$

$$x = 54.6875$$

$$x = 54.7 \text{ L}$$

15. direct proportion

$$\frac{40}{100} = \frac{3.5}{x}$$

$$40x = 100(3.5)$$

$$40x = 350$$

$$x = \frac{350}{40}$$

$$x = 8.8 \text{ A}$$

17. direct proportion

$$\frac{75}{x} = \frac{3\frac{1}{2}}{5} \qquad \left(3\frac{1}{2} = \frac{7}{2}\right)$$

$$3\frac{1}{2}x = 75(5)$$

$$\frac{7}{2}x = 375$$

$$(2)\frac{7}{2}x = (2)375$$

$$7x = 750$$

$$\frac{7x}{7} = \frac{750}{7}$$

$$x = 107.1428571$$

$$x = 107 \text{ lb}$$

19.

$$I = krt$$

$$320 = k(0.08)(4)$$

$$320 = 0.32k$$

$$k = \frac{320}{0.32}$$

$$k = 1,000$$

$$I = 1,000rt$$

$$I = 1,000(0.09)(5)$$

$$I = \$450$$

Chapter Review Exercises

1–5.

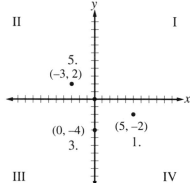

7. The origin has coordinates $(0, 0)$.

9.
A $(3, 0)$
B $(2, 2)$
C $(2, -5)$
D $(-4, -1)$
E $(-3, 1)$

11.

x	y	$2x - 3$
-1	-5	$2(-1) - 3$
1	-1	$2(1) - 3$
3	3	$2(3) - 3$

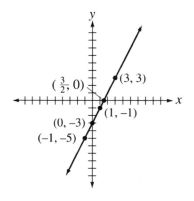

13.

x	y	$3x$
-1	-3	$3(-1)$
0	0	$3(0)$
1	3	$3(1)$

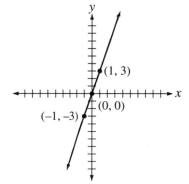

15.

x	y	$-3x$
-1	3	$-3(-1)$
0	0	$-3(0)$
1	-3	$-3(1)$

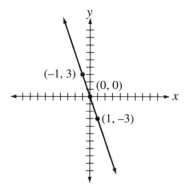

17.

x	y	$2x + 1$
-1	-1	$2(-1) + 1$
0	1	$2(0) + 1$
1	3	$2(1) + 1$

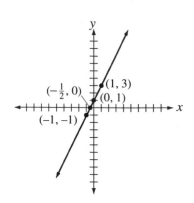

19.

x	y	$4x - 2$
-2	-10	$4(-2) - 2$
-1	-6	$4(-1) - 2$
0	-2	$4(0) - 2$
1	2	$4(1) - 2$
2	6	$4(2) - 2$
3	10	$4(3) - 2$

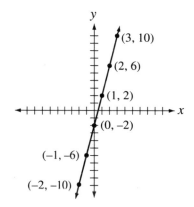

21.

x	y	$\dfrac{1}{2}x - 2$
-2	-3	$\frac{1}{2}(-2) - 2$
-1	$-2\frac{1}{2}$	$\frac{1}{2}(-1) - 2$
0	-2	$\frac{1}{2}(0) - 2$
1	$-1\frac{1}{2}$	$\frac{1}{2}(1) - 2$
2	-1	$\frac{1}{2}(2) - 2$
4	0	$\frac{1}{2}(4) - 2$

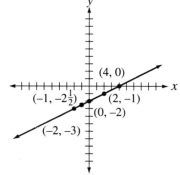

23. $2x - 3y = 12;\quad (-2, -3)$
$$2(-2) - 3(-3) = 12$$
$$-4 + 9 = 12$$
$$5 \neq 12$$
$(-2, -3)$ is not a solution.

25. $2x - 3y = 12;\quad (3, -2)$
$$2(3) - 3(-2) = 12$$
$$6 + 6 = 12$$
$$12 = 12$$
$(3, -2)$ is a solution.

27. $2x - 3y = 12;\quad (6, 0)$
$$2(6) - 3(0) = 12$$
$$12 - 0 = 12$$
$$12 = 12$$
$(6, 0)$ is a solution.

29. $y = 3x - 1$ when $x = -2$
$$y = 3(-2) - 1$$
$$y = -6 - 1$$
$$y = -7$$

31. $x - 3y = 5$ when $x = 8$
$$8 - 3y = 5$$
$$-3y = 5 - 8$$
$$-3y = -3$$
$$y = 1$$

33. $x = -4y - 1$ Intercepts

x	y
0	$-\dfrac{1}{4}$
-1	0

Let $x = 0$.

$x = -4y - 1$
$0 = -4y - 1$
$1 = -4y$
$-\dfrac{1}{4} = y$

Let $y = 0$.

$x = -4y - 1$
$x = -4(0) - 1$
$x = 0 - 1$
$x = -1$

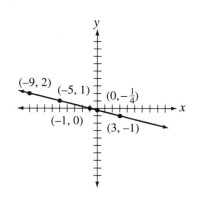

35. $3x - y = 1$ Intercepts

x	y
0	-1
$\dfrac{1}{3}$	0

Let $x = 0$.

$3x - y = 1$
$3(0) - y = 1$
$0 - y = 1$
$-y = 1$
$y = -1$

Let $y = 0$.

$3x - y = 1$
$3x - 0 = 1$
$3x = 1$
$x = \dfrac{1}{3}$

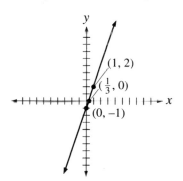

37. $\dfrac{1}{2}x + \dfrac{1}{3}y = 1$ Intercepts

x	y
0	3
2	0

Let $x = 0$.

$\dfrac{1}{2}(0) + \dfrac{1}{3}y = 1$

$0 + \dfrac{1}{3}y = 1$

$3\left[\dfrac{1}{3}y\right] = 1(3)$

$y = 3$

Let $y = 0$.

$\dfrac{1}{2}x + \dfrac{1}{3}(0) = 1$

$\dfrac{1}{2}x = 1$

$2\left[\dfrac{1}{2}x\right] = 1(2)$

$x = 2$

39. $y = 5x - 2$

$m = \dfrac{5}{1}$ $b = -2$

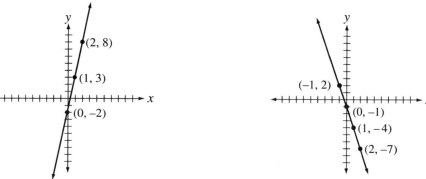

41. $y = -3x - 1$

$m = \dfrac{-3}{1}$ $b = -1$

43. $x - y = 4$

$-y = -x + 4$

$\dfrac{-y}{-1} = \dfrac{-x}{-1} + \dfrac{4}{-1}$

$y = x - 4$

$m = 1 = \dfrac{1}{1} \quad b = -4$

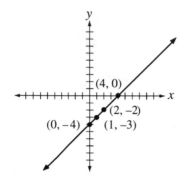

45. $x - 2y = -1$

$-2y = -x - 1$

$\dfrac{-2y}{-2} = \dfrac{-x}{-2} - \dfrac{1}{-2}$

$y = \dfrac{1}{2}x + \dfrac{1}{2}$

$m = \dfrac{1}{2} \qquad b = \dfrac{1}{2}$

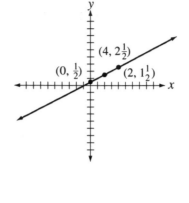

47. $y - 2x = -2$

$y = 2x - 2$

$m = 2 = \dfrac{2}{1} \quad b = -2$

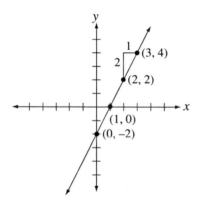

49. $y = 0.5x - 3 \qquad m = 0.5 = \dfrac{1}{2} \qquad b = -3$

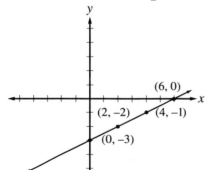

| ZOOM | 6: Z Standard | Y = | CLEAR | · | 5 | X, θ, T, n | – | 3 | GRAPH |

51. $y = 20x - 15$

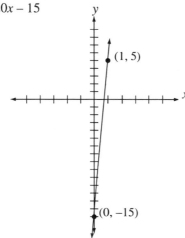

53. $y = 5x + 8{,}000$
$= 5(10{,}000) + 8{,}000$
$= 50{,}000 + 8{,}000$
$y = 58{,}000$
It costs \$58,000 to make 10,000 widgets.

| ZOOM | 6: Z Standard | $Y =$ | CLEAR | 20 | X, θ, T, n | $-$ | 15 | GRAPH |

55.
$$x + 2 = 8$$
$$x + 2 - 8 = 0$$
$$x - 6 = 0$$
$$y = x - 6$$

| $Y =$ | CLEAR |

| X, θ, T, n | -6 | CALC | 2: ZERO |

Set cursor to the left of $y = 0$, $\boxed{\text{ENTER}}$.

Set cursor to the right of $y = 0$, $\boxed{\text{ENTER}}$.

Set cursor near $y = 0$, $\boxed{\text{ENTER}}$.

Read value of x,
$x = 6$

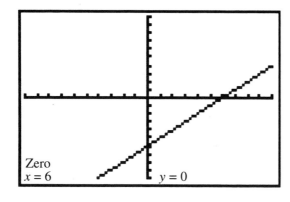

57.
$$2x + 1 = 5x + 7$$
$$0 = 5x - 2x + 7 - 1$$
$$0 = 3x + 6$$
$$y = 3x + 6$$

| $Y =$ | CLEAR |

| 3 | X, θ, T, n | $+$ | 6 |

| CALC | 2: ZERO |

Set cursor to the left of $y = 0$, $\boxed{\text{ENTER}}$.

Set cursor to the right of $y = 0$, $\boxed{\text{ENTER}}$.

Set cursor near $y = 0$, $\boxed{\text{ENTER}}$.

Read value of x,
$x = -2$

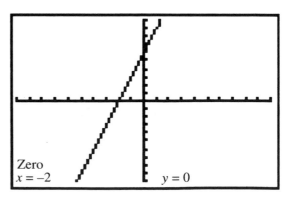

59.
$$5(x + 2) = 3(x + 4)$$
$$5x + 10 = 3x + 12$$
$$5x - 3x + 10 - 12 = 0$$
$$2x - 2 = 0$$
$$y = 2x - 2$$

| $Y =$ | CLEAR |

2 | X, θ, T, n | $-$ | 2

| CALC | | 2: ZERO |

Set cursor to the left of $y = 0$, | ENTER | .

Set cursor to the right of $y = 0$, | ENTER | .

Set cursor near $y = 0$, | ENTER | .

Read value of x,
$x = 1$

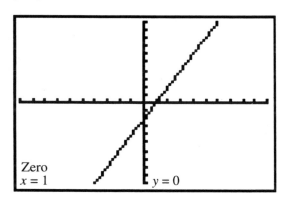

Zero
$x = 1$ $y = 0$

61. $y = 3x + \dfrac{1}{4}$

$m = 3;\ b = \dfrac{1}{4}$

63. $y = -5x + 4$
$m = -5;\ b = 4$

65. $x = 8$
no slope, or undefined slope
no y-intercept

67. $y = \dfrac{x}{8} - 5$

$y = \dfrac{1}{8}x - 5$

$m = \dfrac{1}{8};\ b = -5$

69. $2x + y = 8$
$y = -2x + 8$
$m = -2;\ b = 8$

71. $3x - 2y = 6$
$$-2y = -3x + 6$$
$$\frac{-2y}{-2} = \frac{-3x}{-2} + \frac{6}{-2}$$
$$y = \frac{3}{2}x - 3$$
$$m = \frac{3}{2};\ b = -3$$

73. $\dfrac{3}{5}x - y = 4$
$$-y = -\frac{3}{5}x + 4$$
$$\frac{-y}{-1} = \frac{-\frac{3}{5}x}{-1} + \frac{4}{-1}$$
$$y = \frac{3}{5}x - 4$$
$$m = \frac{3}{5};\ b = -4$$

75. $3y = 5$
$$\frac{3y}{3} = \frac{5}{3}$$
$$y = \frac{5}{3} \text{ or } y = 0x + \frac{5}{3}$$
$$m = 0;\ b = \frac{5}{3}$$

77. $(-2, 2)$ and $(1, 3)$
$$m = \frac{y_2 - y_1}{x_2 - x_1} = \frac{3 - 2}{1 - (-2)} = \frac{1}{3}$$

79. $(3, 2)$ and $(5, 6)$
$$m = \frac{y_2 - y_1}{x_2 - x_1} = \frac{6 - 2}{5 - 3} = \frac{4}{2} = 2$$

81. $(4, 3)$ and $(-4, -2)$
$$m = \frac{y_2 - y_1}{x_2 - x_1} = \frac{-2 - 3}{-4 - 4} = \frac{-5}{-8} = \frac{5}{8}$$

83. $(3, -4)$ and $(0, 0)$

$$m = \frac{y_2 - y_1}{x_2 - x_1} = \frac{0 - (-4)}{0 - 3} = \frac{4}{-3} = -\frac{4}{3}$$

85. $(-4, 1)$ and $(-4, 3)$

$$m = \frac{y_2 - y_1}{x_2 - x_1} = \frac{3 - 1}{-4 - (-4)} = \frac{2}{0}; \text{ undefined slope}$$

87. $(5, 0)$ and $(-2, 4)$

$$m = \frac{y_2 - y_1}{x_2 - x_1} = \frac{4 - 0}{-2 - 5} = \frac{4}{-7} = -\frac{4}{7}$$

89. $(-4, -8)$ and $(-2, -1)$

$$m = \frac{y_2 - y_1}{x_2 - x_1} = \frac{(-1) - (-8)}{(-2) - (-4)} = \frac{7}{2}$$

91. $(5, -3)$ and $(-1, -3)$

$$m = \frac{y_2 - y_1}{x_2 - x_1} = \frac{(-3) - (-3)}{(-1) - 5} = \frac{0}{-6} = 0$$

93. $(-7, 0)$ and $(-7, 5)$

$$m = \frac{y_2 - y_1}{x_2 - x_1} = \frac{5 - 0}{(-7) - (-7)} = \frac{5}{0}; \text{ undefined slope}$$

95. points on the same horizontal line (slope $m = 0$; thus y_1 and y_2 are same value)
$(3, 5)$ and $(4, 5)$ Answers will vary.

97. $(1998, 1{,}791)$ $(1999, 1{,}847)$

$$m = \frac{y_2 - y_1}{x_2 - x_1}$$

$$\text{rate of change} = \frac{1{,}847 - 1{,}791}{1999 - 1998} = \frac{56}{1} = 56$$

Rate of change was an *increase* of \$56.

99. $(1993, 1{,}613)$ $(1994, 1{,}650)$

$$m = \frac{y_2 - y_1}{x_2 - x_1}$$

$$\text{rate of change} = \frac{1{,}650 - 1{,}613}{1994 - 1993} = \frac{37}{1} = 37$$

Rate of change = \$37

101. Explain why the rate of change is different in Exercises 98, 99, and 100.
The table of values does not represent a perfect linear function, so the rate of change (slope) varies.

103. $(-6, 2)$ and $m = \dfrac{1}{3}$

$$y - y_1 = m(x - x_1)$$
$$y - 2 = \frac{1}{3}(x - (-6))$$
$$y - 2 = \frac{1}{3}(x + 6)$$
$$y - 2 = \frac{1}{3}x + 2$$
$$y = \frac{1}{3}x + 2 + 2$$
$$y = \frac{1}{3}x + 4$$

105. $(4, 0)$ and $m = \dfrac{3}{4}$

$$y - y_1 = m(x - x_1)$$
$$y - 0 = \frac{3}{4}(x - 4)$$
$$y = \frac{3}{4}x - 3$$

107. $(2, 3)$ and $m = 4$

$$y - y_1 = m(x - x_1)$$
$$y - 3 = 4(x - 2)$$
$$y - 3 = 4x - 8$$
$$y = 4x - 8 + 3$$
$$y = 4x - 5$$

109. $(-5, 2)$ and $(6, 1)$

$$m = \frac{y_2 - y_1}{x_2 - x_1} = \frac{1 - 2}{6 - (-5)} = \frac{-1}{11}$$

$$y - y_1 = m(x - x_1)$$

$$y - 2 = \frac{-1}{11}(x - (-5))$$

$$y - 2 = \frac{-1}{11}(x + 5)$$

$$y - 2 = \frac{-1}{11}x - \frac{5}{11}$$

$$y = \frac{-1}{11}x - \frac{5}{11} + 2$$

$$y = \frac{-1}{11}x - \frac{5}{11} + \frac{22}{11}$$

$$y = -\frac{1}{11}x + \frac{17}{11}$$

111. $(-1, -3)$ and $(3, 4)$

$$m = \frac{y_2 - y_1}{x_2 - x_1} = \frac{4 - (-3)}{3 - (-1)} = \frac{7}{4}$$

$$y - y_1 = m(x - x_1)$$

$$y - (-3) = \frac{7}{4}(x - (-1))$$

$$y + 3 = \frac{7}{4}(x + 1)$$

$$y + 3 = \frac{7}{4}x + \frac{7}{4}$$

$$y = \frac{7}{4}x + \frac{7}{4} - 3$$

$$y = \frac{7}{4}x + \frac{7}{4} - \frac{12}{4}$$

$$y = \frac{7}{4}x - \frac{5}{4}$$

113. $(-2, -3)$ and $(3, 6)$

$$m = \frac{y_2 - y_1}{x_2 - x_1} = \frac{6 - (-3)}{3 - (-2)} = \frac{9}{5}$$

$$y - y_1 = m(x - x_1)$$

$$y - (-3) = \frac{9}{5}(x - (-2))$$

$$y + 3 = \frac{9}{5}(x + 2)$$

$$y + 3 = \frac{9}{5}x + \frac{18}{5}$$

$$y = \frac{9}{5}x + \frac{18}{5} - 3$$

$$y = \frac{9}{5}x + \frac{18}{5} - \frac{15}{5}$$

$$y = \frac{9}{5}x + \frac{3}{5}$$

115. $(5, 2)$ and $(6, 3)$

$$m = \frac{y_2 - y_1}{x_2 - x_1} = \frac{3 - 2}{6 - 5} = \frac{1}{1} = 1$$

$$y - y_1 = m(x - x_1)$$

$$y - 2 = 1(x - 5)$$

$$y - 2 = x - 5$$

$$y = x - 5 + 2$$

$$y = x - 3$$

117. $(-1, -2)$ and $(-3, -4)$

$$m = \frac{y_2 - y_1}{x_2 - x_1} = \frac{-4 - (-2)}{-3 - (-1)} = \frac{-2}{-2} = 1$$

$$y - y_1 = m(x - x_1)$$

$$y - (-2) = 1(x - (-1))$$

$$y + 2 = 1(x + 1)$$

$$y + 2 = x + 1$$

$$y = x + 1 - 2$$

$$y = x - 1$$

119. $(5, -2)$ and $(3, -2)$

$$m = \frac{y_2 - y_1}{x_2 - x_1} = \frac{-2 - (-2)}{3 - 5} = \frac{0}{-2} = 0$$

horizontal line $y = -2$

121. $(80, \$3{,}800)$
$(120, \$4{,}200)$

$$m = \frac{y_2 - y_1}{x_2 - x_1}$$

$$m = \frac{4{,}200 - 3{,}800}{120 - 80} = \frac{400}{40}$$

$$m = \frac{10}{1} = 10$$

$$y - 3{,}800 = 10(x - 80)$$

$$y - 3{,}800 = 10x - 800$$

$$y = 10x - 800 + 3{,}800$$

$$y = 10x + 3{,}000$$

$$S = 10x + 3{,}000$$

123. $m = 3 \quad b = -2$

$$y = mx + b$$

$$y = 3x - 2$$

125. $m = \frac{2}{1} = 2 \quad b = -2$

$$y = mx + b$$

$$y = 2x - 2$$

127. parallel to $x + y = 4$

$$y = -x + 4$$

$m_{given} = -1,\ m_{parallel} = -1$

$$y - y_1 = m(x - x_1) \qquad \text{point } (2, 5)$$
$$y - 5 = -1(x - 2)$$
$$y - 5 = -1x + 2$$
$$y = -1x + 2 + 5$$
$$y = -1x + 7$$

standard form: $x + y = 7$

129. parallel to $2y = x - 3$

$$\frac{2y}{2} = \frac{x}{2} - \frac{3}{2}$$
$$y = \frac{1}{2}x - \frac{3}{2}$$

$m_{given} = \dfrac{1}{2},\ m_{parallel} = \dfrac{1}{2}$

$$y - y_1 = m(x - x_1) \qquad \text{point } (2, -3)$$
$$y - (-3) = \frac{1}{2}(x - 2)$$
$$y + 3 = \frac{1}{2}x - 1$$
$$y = \frac{1}{2}x - 1 - 3$$
$$y = \frac{1}{2}x - 4$$
$$(2)y = (2)\frac{1}{2}x - (2)4$$
$$2y = x - 8$$
$$-x + 2y = -8$$

standard form: $x - 2y = 8$

131. parallel to $x - 3y = 5$

$$-3y = -x + 5$$
$$\frac{-3y}{-3} = \frac{-x}{-3} + \frac{5}{-3}$$
$$y = \frac{1}{3}x - \frac{5}{3}$$

$m_{given} = \dfrac{1}{3},\ m_{parallel} = \dfrac{1}{3}$

$$y - y_1 = m(x - x_1) \qquad \text{point } (5, -5)$$
$$y - (-5) = \frac{1}{3}(x - 5)$$
$$y + 5 = \frac{1}{3}x - \frac{5}{3}$$
$$y = \frac{1}{3}x - \frac{5}{3} - 5$$
$$y = \frac{1}{3}x - \frac{5}{3} - \frac{15}{3}$$
$$y = \frac{1}{3}x - \frac{20}{3}$$
$$(3)y = (\not{3})\frac{1}{\not{3}}x - (\not{3})\frac{20}{\not{3}}$$
$$3y = x - 20$$
$$-x + 3y = -20$$

standard form: $x - 3y = 20$

133. parallel to $x + 3y = 6$

$$3y = -x + 6$$
$$\frac{3y}{3} = \frac{-x}{3} + \frac{6}{3}$$
$$y = \frac{-1}{3}x + 2$$

$m_{given} = \dfrac{-1}{3},\ m_{parallel} = \dfrac{-1}{3}$

$$y - y_1 = m(x - x_1) \qquad \text{point } (-4, -2)$$
$$y - (-2) = \frac{-1}{3}(x - (-4))$$
$$y + 2 = \frac{-1}{3}x - \frac{4}{3}$$
$$y = \frac{-1}{3}x - \frac{4}{3} - \frac{6}{3}$$
$$y = \frac{-1}{3}x - \frac{10}{3}$$
$$(3)y = (\not{3})\frac{-1}{\not{3}}x - (\not{3})\frac{10}{\not{3}}$$
$$3y = -x - 10$$

standard form: $x + 3y = -10$

135. parallel to $3x - 4y = 0$

$$-4y = -3x$$
$$\frac{-4y}{-4} = \frac{-3x}{-4}$$
$$y = \frac{3}{4}x$$

$$m_{given} = \frac{3}{4}, \ m_{parallel} = \frac{3}{4}$$

$$y - y_1 = m(x - x_1) \qquad point \left(\frac{1}{3}, 2\right)$$

$$y - 2 = \frac{3}{4}\left(x - \frac{1}{3}\right)$$

$$y - 2 = \frac{3}{4}x - \frac{1}{4}$$

$$y = \frac{3}{4}x - \frac{1}{4} + 2$$

$$y = \frac{3}{4}x - \frac{1}{4} + \frac{8}{4}$$

$$y = \frac{3}{4}x + \frac{7}{4}$$

$$(4)y = (\cancel{4})\frac{3}{\cancel{4}}x + (\cancel{4})\frac{7}{\cancel{4}}$$

$$4y = 3x + 7$$
$$-3x + 4y = 7$$

standard form: $3x - 4y = -7$

141. perpendicular to $5x + y = 8$

$$y = -5x + 8$$
$$m_{given} = -5, \ m_{perpendicular} = +\frac{1}{5}$$

$$y - y_1 = m(x - x_1) \qquad point (-1, 2)$$

$$y - 2 = \frac{1}{5}(x - (-1))$$

$$y - 2 = \frac{1}{5}x + \frac{1}{5}$$

$$y = \frac{1}{5}x + \frac{1}{5} + 2$$

$$y = \frac{1}{5}x + \frac{1}{5} + \frac{10}{5}$$

$$y = \frac{1}{5}x + \frac{11}{5}$$

$$(5)y = (\cancel{5})\frac{1}{\cancel{5}}x - (\cancel{5})\frac{11}{\cancel{5}}$$

$$5y = x + 11$$
$$-x + 5y = 11$$

standard form: $x - 5y = -11$

137. perpendicular to $x + y = 4$

$$y = -x + 4$$
$$m_{given} = -1, \ m_{perpendicular} = 1$$

$$y - y_1 = m(x - x_1) \qquad point (-3, 1)$$

$$y - 1 = 1(x - (-3))$$
$$y - 1 = 1x + 3$$
$$y = 1x + 3 + 1$$
$$y = x + 4$$
$$-x + y = 4$$

standard form: $x - y = -4$

139. perpendicular to $x + 2y = 5$

$$2y = -x + 5$$
$$\frac{2y}{2} = \frac{-x}{2} + \frac{5}{2}$$
$$y = \frac{-1}{2}x + \frac{5}{2}$$

$$m_{given} = \frac{-1}{2}, \ m_{perpendicular} = +2$$

$$y - y_1 = m(x - x_1) \qquad point (-2, 0)$$
$$y - 0 = 2(x - (-2))$$
$$y = 2x + 4$$
$$-2x + y = 4$$

standard form: $2x - y = -4$

143. perpendicular to $5x - y = 10$

$$-y = -5x + 10$$
$$\frac{-y}{-} = \frac{-5x}{-1} + \frac{10}{-1}$$
$$y = 5x - 10$$

$$m_{given} = 5, \ m_{perpendicular} = \frac{-1}{5}$$

$$y - y_1 = m(x - x_1) \qquad point \left(\frac{1}{2}, 3\right)$$

$$y - 3 = \frac{-1}{5}\left(x - \frac{1}{2}\right)$$

$$y - 3 = \frac{-1}{5}x + \frac{1}{10}$$

$$y = \frac{-1}{5}x + \frac{1}{10} + 3$$

$$y = \frac{-1}{5}x + \frac{1}{10} + \frac{30}{10}$$

$$y = \frac{-1}{5}x + \frac{31}{10}$$

$$(10)y = (10)\frac{-1}{5}x + (10)\frac{31}{10}$$

$$10y = -2x + 31$$

standard form: $2x + 10y = 31$

145. perpendicular to $4x - y = 8$

$$-y = -4x + 8$$

$$\frac{-y}{-1} = \frac{-4x}{-1} + \frac{8}{-1}$$

$$y = 4x - 8$$

$$m_{\text{given}} = 4, \ m_{\text{perpendicular}} = \frac{-1}{4}$$

$$y - y_1 = m(x - x_1) \quad \text{point}\left(4, -\frac{1}{2}\right)$$

$$y - \frac{-1}{2} = \frac{-1}{4}(x - 4)$$

$$y + \frac{1}{2} = \frac{-1}{4}x + 1$$

$$y = \frac{-1}{4}x + 1 - \frac{1}{2}$$

$$y = \frac{-1}{4}x + \frac{2}{2} - \frac{1}{2}$$

$$y = \frac{-1}{4}x + \frac{1}{2}$$

$$(4)y = (4)\frac{-1}{4}x + (4)\frac{1}{2}$$

$$4y = -x + 2$$

standard form: $x + 4y = 2$

Chapter 9 Practice Test

1.

x	y	$\frac{1}{2}x$
-2	-1	$\frac{1}{2}(-2)$
0	0	$\frac{1}{2}(0)$
2	1	$\frac{1}{2}(2)$

3.

x	y	$2x - 4$
-1	-6	$2(-1) - 4$
0	-4	$2(0) - 4$
1	-2	$2(1) - 4$
2	0	$2(2) - 4$

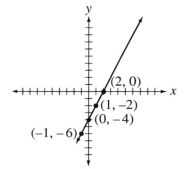

5.

$$3x + 2 = 5$$
$$3x + 2 - 5 = 0$$
$$3x - 3 = 0$$
$$y = 3x - 3$$

 CLEAR

3 $\boxed{X, \theta, T, n}$ $\boxed{-}$ 3

$\boxed{\text{CALC}}$ $\boxed{\text{2: ZERO}}$

Set cursor to the left of $y = 0$, $\boxed{\text{ENTER}}$.

Set cursor to the right of $y = 0$, $\boxed{\text{ENTER}}$.

Set cursor near $y = 0$, $\boxed{\text{ENTER}}$.

Read the value of x,
$x = 1$

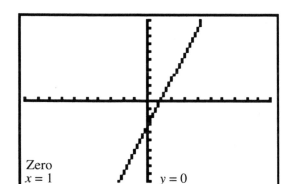

7.

$$x + y = 7 \qquad \text{if } x = 2$$
$$2 + y = 7$$
$$y = 7 - 2$$
$$y = 5$$

Solution: (2, 5)

9.

$$y = 3x - 800$$
$$y = 3(8,000) - 800$$
$$= 24,000 - 800$$
$$y = \$23,200$$

11. $x + y = -5$

x-intercept	y-intercept
$y = 0$	$x = 0$
$x + 0 = -5$	$0 + y = -5$
$x = -5$	$y = -5$
$(-5, 0)$	$(0, -5)$

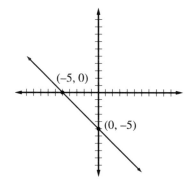

13. $x + 2y = 8$

x-intercept	y-intercept
$y = 0$	$x = 0$
$x + 2(0) = 8$	$0 + 2y = 8$
$x + 0 = 8$	$2y = 8$
$x = 8$	$y = 4$
$(8, 0)$	$(0, 4)$

15.

$$2x + y = -3$$
$$y = -2x - 3$$
$$m = \frac{-2}{+1} \qquad y\text{-intercept} = (0, -3)$$

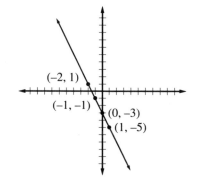

17. $(-3, 6)$ and $(3, 2)$

$$m = \frac{y_2 - y_1}{x_2 - x_1} = \frac{2 - 6}{3 - (-3)} = \frac{-4}{6} = -\frac{2}{3}$$

19. $-2x + y = 34$
$y = 2x + 34$
$m = 2;\ b = 34$

21. $x = 4y$
$4y = x$
$$\frac{4y}{4} = \frac{x}{4}$$
$$y = \frac{1}{4}x$$
$$m = \frac{1}{4};\ b = 0$$

23. $(3, -5), \qquad m = \frac{2}{3}$

$y - y_1 = m(x - x_1)$

$y - (-5) = \frac{2}{3}(x - 3)$

$y + 5 = \frac{2}{3}x - 2$

$y = \frac{2}{3}x - 2 - 5$

$y = \frac{2}{3}x - 7$

25. $(1, 3)$ and $(4, 5)$

$$m = \frac{y_2 - y_1}{x_2 - x_1} = \frac{5 - 3}{4 - 1} = \frac{2}{3}$$

$y - y_1 = m(x - x_1)$

$y - 3 = \frac{2}{3}(x - 1)$

$y - 3 = \frac{2}{3}x - \frac{2}{3}$

$y = \frac{2}{3}x - \frac{2}{3} + 3$

$y = \frac{2}{3}x - \frac{2}{3} + \frac{9}{3}$

$y = \frac{2}{3}x + \frac{7}{3}$

27. $(5, 2)$ and $(-1, 2)$

$$m = \frac{y_2 - y_1}{x_2 - x_1} = \frac{2 - 2}{-1 - 5} = \frac{0}{-6} = 0$$

horizontal line

$y = 2$

29. $m = \frac{3}{2};\ b = 3$

$y = mx + b$

$y = \frac{3}{2}x + 3$

31. parallel to $2x + y = 4$

$\qquad\qquad y = -2x + 4$

$m_{\text{given}} = -2,\ m_{\text{parallel}} = -2$

$y - y_1 = m(x - x_1) \qquad$ point $(4, -3)$

$y - (-3) = -2(x - 4)$

$y + 3 = -2(x - 4)$

$y + 3 = -2x + 8$

$y = -2x + 8 - 3$

$y = -2x + 5$

standard form: $\qquad 2x + y = 5$

33. perpendicular to $2x + y = 4$

$\qquad\qquad y = -2x + 4$

$m_{\text{given}} = -2,\ m_{\text{perpendicular}} = \frac{1}{2}$

$y - y_1 = m(x - x_1) \qquad$ point $(4, -3)$

$y - (-3) = \frac{1}{2}(x - 4)$

$y + 3 = \frac{1}{2}x - 2$

$y = \frac{1}{2}x - 2 - 3$

$y = \frac{1}{2}x - 5$

$(2)y = (2)\frac{1}{2}x - (2)5$

$2y = x - 10$

$-x + 2y = -10$

standard form: $x - 2y = 10$

Chapter Review Exercises

1. $y = 5x - 1$
$y = 3x + 3$ Solution $(2, 9)$

$y = 5x - 1$ $y = 3x + 3$

$m = 5$ or $\dfrac{5}{1}$; $b = -1$ $m = 3$ or $\dfrac{3}{1}$; $b = 3$

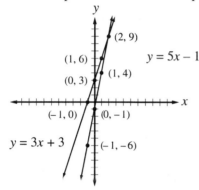

3. $y = 5x + 1$
$y = 3x - 1$ Solution $(-1, -4)$

$y = 5x + 1$ $y = 3x - 1$

$m = 5$ or $\dfrac{5}{1}$; $b = 1$ $m = 3$ or $\dfrac{3}{1}$; $b = -1$

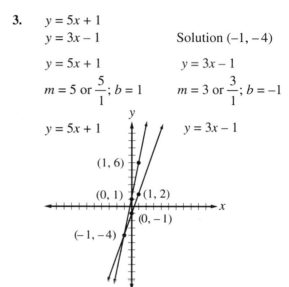

5. $x + y = 8$
$x - y = 2$ Solution $(5, 3)$

$x + y = 8$ $x - y = 2$
$\quad y = -x + 8$ $-y = -x + 2$

$m = \dfrac{-1}{1}$; $b = 8$ $\dfrac{-y}{-1} = \dfrac{-x}{-1} + \dfrac{2}{-1}$

                  $y = x - 2$

              $m = \dfrac{1}{1}$; $b = -2$

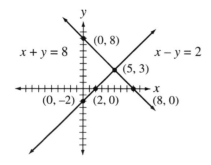

7. $2x + 2y = 10$
$3x + 3y = 15$

$2x + 2y = 10$
$\quad 2y = -2x + 10$
$\quad \dfrac{2y}{2} = \dfrac{-2x}{2} + \dfrac{10}{2}$
$\quad\quad y = -1x + 5$

$m = \dfrac{-1}{1}$; $b = 5$

Any ordered pair that is a solution of one equation is a solution of the other equation. There are many solutions.
Dependent; lines coincide

$3x + 3y = 15$
$\quad 3y = -3x + 15$
$\quad \dfrac{3y}{3} = \dfrac{-3x}{3} + \dfrac{15}{3}$
$\quad\quad y = -1x + 5$
$\quad\quad y = \dfrac{-1}{1}$; $b = 5$

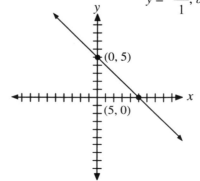

9.

$$3x + y = 9$$
$$\underline{2x - y = 6}$$
$$5x \quad = 15$$
$$\frac{5x}{5} = \frac{15}{5}$$
$$x = 3$$

$$3x + y = 9$$
$$3(3) + y = 9$$
$$9 + y = 9$$
$$y = 9 - 9$$
$$y = 0$$
$$x = 3, y = 0$$
$$(3, 0)$$

11.

$$5x - 3y = 8$$
$$\underline{-5x + 2y = -7}$$
$$-y = 1$$
$$y = -1$$
$$5x - 3y = 8$$
$$5x - 3(-1) = 8$$
$$5x + 3 = 8$$
$$5x = 8 - 3$$
$$5x = 5$$
$$x = 1$$
$$x = 1, y = -1$$
$$(1, -1)$$

13.

$$Q = 2P + 8$$
$$2Q + 3P = 2$$
$$-2(Q - 2P) = (8)(-2)$$
$$2Q + 3P = 2$$
$$-2Q + 4P = -16$$
$$\underline{2Q + 3P = \quad 2}$$
$$7P = -14$$
$$P = -2$$

$$Q = 2P + 8$$
$$Q = 2(-2) + 8$$
$$Q = -4 + 8$$
$$Q = 4$$
$$P = -2, Q = 4$$
$$(-2, 4)$$

15.

$$r = 2y + 6$$
$$2r + y = 2$$
$$r - 2y = 6$$
$$2(2r + y) = (2)(2)$$
$$r - 2y = 6$$
$$\underline{4r + 2y = 4}$$
$$5r \quad = 10$$
$$r = 2$$
$$r = 2y + 6$$
$$2 = 2y + 6$$
$$-4 = 2y$$
$$-2 = y$$
$$r = 2, y = -2$$
$$(2, -2)$$

17.

$$c = 2y$$
$$2c + 3y = 21$$
$$-2(c - 2y) = (0)(-2)$$
$$2c + 3y = 21$$
$$-2c + 4y = 0$$
$$\underline{2c + 3y = 21}$$
$$7y = 21$$
$$y = 3$$
$$c = 2y$$
$$c = 2(3)$$
$$c = 6$$
$$c = 6, y = 3$$
$$(6, 3)$$

19.

$$3R - 2S = 7$$
$$-14 = -6R + 4S$$
$$-2(3R - 2S) = (7)(-2)$$
$$6R - 4S = 14$$
$$-6R + 4S = -14$$
$$\underline{6R - 4S = 14}$$
$$0 = 0 \text{ true}$$

Dependent
Any ordered pair that is a solution of one equation is a solution of the other equation. There are many solutions.

21.

$$c = 2 + 3d$$
$$3c - 14 = d$$
$$-3(c - 3d) = (2)(-3)$$
$$3c - d = 14$$
$$-3c + 9d = -6$$
$$\underline{3c - d = 14}$$
$$8d = 8$$
$$d = 1$$
$$c = 2 + 3d$$
$$c = 2 + 3(1)$$
$$c = 2 + 3$$
$$c = 5$$
$$c = 5, d = 1$$
$$(5, 1)$$

23.

$$x - 18 = -6y$$
$$4x - 0 = 3y$$
$$x + 6y = 18$$
$$2(4x - 3y) = (0)(2)$$
$$x + 6y = 18$$
$$\underline{8x - 6y = 0}$$
$$9x \quad = 18$$
$$x = 2$$
$$x - 18 = -6y$$
$$2 - 18 = -6y$$
$$-16 = -6y$$
$$\frac{-16}{-6} = \frac{-6y}{-6}$$
$$\frac{8}{3} = y$$
$$x = 2, y = \frac{8}{3}$$
$$\left(2, \frac{8}{3}\right)$$

25.

$$3a - 2b = 6$$
$$6a - 12 = b$$
$$3a - 2b = 6$$
$$-2(6a - b) = (12)(-2)$$
$$3a - 2b = \quad 6$$
$$\underline{-12a + 2b = -24}$$
$$-9a \quad = -18$$
$$a = 2$$
$$3a - 2b = 6$$
$$3(2) - 2b = 6$$
$$6 - 2b = 6$$
$$-2b = 0$$
$$b = 0$$
$$a = 2, b = 0$$
$$(2, 0)$$

27. $x + 2y = 7$
$x - y = 1$

$x + 2y = 7$
$2(x - y) = (1)(2)$

$x + 2y = 7$
$2x - 2y = 2$
$\overline{3x = 9}$
$x = 3$

$x + 2y = 7$
$3 + 2y = 7$
$2y = 4$
$y = 2$
$x = 3, y = 2$
$(3, 2)$

29. $x + 2r = 5.5$
$2x = 1.5r$

$-2(x + 2r) = 5.5(-2)$
$2x - 1.5r = 0$

$-2x - 4r = -11.0$
$2x - 1.5r = 0$
$\overline{-5.5r = -11.0}$
$r = \dfrac{-11.0}{-5.5}$
$r = 2$

$2x = 1.5r$
$2x = 1.5(2)$
$2x = 3$
$x = \dfrac{3}{2}$

$\left(2, \dfrac{3}{2}\right)$ or

$(2, 1.5)$

31. $a + 7b = 32$
$3a - b = 8$

$a + 7b = 32$
$a = 32 - 7b$

$3a - b = 8$
$3(32 - 7b) - b = 8$
$96 - 21b - b = 8$
$96 - 22b = 8$
$-22b = -88$
$b = 4$

$a + 7b = 32$
$a + 7(4) = 32$
$a + 28 = 32$
$a = 4$
$a = 4, b = 4$
$(4, 4)$

33. $c - d = 2$
$c = 12 - d$

$c - d = 2$
$(12 - d) - d = 2$
$12 - d - d = 2$
$12 - 2d = 2$
$-2d = -10$
$d = 5$

$c = 12 - d$
$c = 12 - 5$
$c = 7$
$c = 7, d = 5$
$(7, 5)$

35. $7x - 4 = -4y$
$3x + y = 6$

$3x + y = 6$
$y = 6 - 3x$

$7x - 4 = -4y$
$7x - 4 = -4(6 - 3x)$
$7x - 4 = -24 + 12x$
$-5x = -20$
$x = 4$

$7x - 4 = -4y$
$7(4) - 4 = -4y$
$28 - 4 = -4y$
$24 = -4y$
$-6 = y$
$x = 4, \ y = -6$
$(4, -6)$

37. $a = 2b + 11$
$3a + 11 = -5b$

$3a + 11 = -5b$
$3(2b + 11) + 11 = -5b$
$6b + 33 + 11 = -5b$
$44 = -11b$
$-4 = b$

$a = 2b + 11$
$a = 2(-4) + 11$
$a = -8 + 11$
$a = 3$
$a = 3, b = -4$
$(3, -4)$

39.
$$c = 2q$$
$$2c + q = 2$$

$$2c + q = 2$$
$$2(2q) + q = 2$$
$$4q + q = 2$$
$$5q = 2$$
$$q = \frac{2}{5}$$

$$c = 2q$$
$$c = 2\left(\frac{2}{5}\right)$$
$$c = \frac{4}{5}$$
$$c = \frac{4}{5}, \ q = \frac{2}{5}$$
$$\left(\frac{4}{5}, \frac{2}{5}\right)$$

41.
$$4x - 2.5y = 2$$
$$2x - 1.5y = -10$$

$$2x - 1.5y = -10$$
$$2x = 1.5y - 10$$
$$\frac{2x}{2} = \frac{1.5y}{2} - \frac{10}{2}$$
$$x = 0.75y - 5$$

$$4x - 2.5y = 2$$
$$4(0.75y - 5) - 2.5y = 2$$
$$3y - 20 - 2.5y = 2$$
$$0.5y = 22$$
$$y = 44$$

$$4x - 2.5y = 2$$
$$4x - 2.5(44) = 2$$
$$4x - 110 = 2$$
$$4x = 112$$
$$x = 28$$
$$x = 28, \ y = 44$$
$$(28, 44)$$

43.
$$4d - 7 = -c$$
$$3c - 6 = -6d$$

$$4d - 7 = -c$$
$$\frac{4d}{-1} - \frac{7}{-1} = \frac{-c}{-1}$$
$$-4d + 7 = c$$

$$3c - 6 = -6d$$
$$3(-4d + 7) - 6 = -6d$$
$$-12d + 21 - 6 = -6d$$
$$-12d + 15 = -6d$$
$$15 = 6d$$
$$\frac{15}{6} = d$$
$$d = \frac{5}{2}$$

$$4d - 7 = -c$$
$$4\left(\frac{5}{2}\right) - 7 = -c$$
$$10 - 7 = -c$$
$$3 = -c$$
$$c = -3$$
$$c = -3, \ d = \frac{5}{2}$$
$$\left(-3, \frac{5}{2}\right)$$

45.
$$3.5a + 2b = 2$$
$$0.5b = 3 - 1.5a$$

$$3.5a + 2b = 2$$
$$2b = -3.5a + 2$$
$$\frac{2b}{2} = \frac{-3.5a}{2} + \frac{2}{2}$$
$$b = -1.75a + 1$$

$$0.5b = 3 - 1.5a$$
$$0.5(-1.75a + 1) = 3 - 1.5a$$
$$-0.875a + 0.5 = 3 - 1.5a$$
$$0.625a = 2.5$$
$$a = 4$$

$$3.5a + 2b = 2$$
$$3.5(4) + 2b = 2$$
$$14 + 2b = 2$$
$$2b = -12$$
$$b = -6$$
$$a = 4, \ b = -6$$
$$(4, -6)$$

47.
$$x + 4y = 20$$
$$4x + 5y = 58$$

$$x + 4y = 20$$
$$x = 20 - 4y$$

$$4x + 5y = 58$$
$$4(20 - 4y) + 5y = 58$$
$$80 - 16y + 5y = 58$$
$$80 - 11y = 58$$
$$-11y = -22$$
$$y = 2$$

$$x + 4y = 20$$
$$x + 4(2) = 20$$
$$x + 8 = 20$$
$$x = 12$$
$$x = 12, \ y = 2$$
$$(12, 2)$$

49.
$$3a + 1 = -2b$$
$$4b + 23 = 15a$$
Use addition method.
$$-2(3a + 2b) = (-1)(-2)$$
$$-15a + 4b = -23$$

$$-6a - 4b = 2$$
$$-15a + 4b = -23$$
$$-21a \quad\quad = -21$$
$$a = 1$$

$$3a + 1 = -2b$$
$$3(1) + 1 = -2b$$
$$3 + 1 = -2b$$
$$\frac{4}{-2} = \frac{-2b}{-2}$$
$$-2 = b$$
$$a = 1, \ b = -2$$
$$(1, -2)$$

51. let x = pay per electrician
y = pay per apprentice

$3x + 4y = 365$
$x + 2y = 145$

$3x + 4y = 365$
$-3(x + 2y = 145)$

$3x + 4y = 365$
$-3x - 6y = -435$
$\overline{-2y = -70}$
$y = 35$

$x + 2y = 145$
$x + 2(35) = 145$
$x + 70 = 145$
$x = 75$
electrician pay = \$75
apprentice pay = \$35

53. let x = cost of each quart of shellac
y = cost of each quart of thinner

$2x + 5y = 22.50$
$3x + 2y = 14.50$

$3(2x + 5y) = (22.50)(3)$
$-2(3x + 2y) = (14.50)(-2)$

$6x + 15y = 67.50$
$-6x - 4y = -29.00$
$\overline{11y = 38.50}$
$y = 3.50$

$2x + 5y = 22.50$
$2x + 5(3.50) = 22.50$
$2x + 17.50 = 22.50$
$2x = 5.00$
$x = 2.50$
cost of shellac = \$2.50
cost of thinner = \$3.50

55. let x = angle 1
y = angle 2

$x + y = 175$
$x - y = 63$

$x + y = 175$
$x - y = 63$
$\overline{2x = 238}$
$x = 119$

$x + y = 175$
$119 + y = 175$
$y = 56$

larger angle = $119°$
smaller angle = $56°$

57. let x = investment at 5%
y = investment at 6%
$x + y = 5,000$
$0.05x + 0.06y = 280$

$x + y = 5,000$
$x = 5,000 - y$

$0.05x + 0.06y = 280$
$0.05(5,000 - y) + 0.06y = 280$
$250 - 0.05y + 0.06y = 280$
$0.01y = 30$
$y = 3,000$

$x + y = 5,000$
$x + 3,000 = 5,000$
$x = 2,000$
amount at 5% = \$2,000
amount at 6% = \$3,000

59. let x = cost per pound of Colombian coffee
y = cost per pound of blended coffee

$$30x + 10y = 190$$
$$20x + 5y = 120$$

$$30x + 10y = 190$$
$$-2(20x + 5y) = (120)(-2)$$

$$30x + 10y = 190$$
$$-40x - 10y = -240$$
$$\overline{}$$
$$-10x = -50$$
$$x = 5$$

$$30x + 10y = 190$$
$$30(5) + 10y = 190$$
$$150 + 10y = 190$$
$$10y = 40$$
$$y = 4$$

cost of Colombian coffee = $5
cost of blended coffee = $4

61. let x = cost of Ohio map
y = cost of Alaska map

$$25x + 8y = 65.55$$
$$20x + 5y = 49.50$$

$$-5(25x + 8y) = (65.55)(-5)$$
$$8(20x + 5y) = (49.50)(8)$$

$$-125x - 40y = -327.75$$
$$160x + 40y = 396.00$$
$$\overline{}$$
$$35x = 68.25$$
$$x = 1.95$$

$$25(1.95) + 8y = 65.55$$
$$48.75 + 8y = 65.55$$
$$8y = 16.80$$
$$y = 2.10$$

cost of Ohio map = $1.95
cost of Alaska map = $2.10

63. let x = name brand suit cost
y = generic label suit cost

$$20x + 35y = 12{,}525$$
$$30x + 35y = 15{,}725$$

$$-1(20x + 35y) = (12{,}525)(-1)$$
$$30x + 35y = 15{,}725$$

$$-20x - 35y = -12{,}525$$
$$30x + 35y = 15{,}725$$
$$\overline{}$$
$$10x = 3{,}200$$
$$x = 320$$

$$20(320) + 35y = 12{,}525$$
$$6{,}400 + 35y = 12{,}525$$
$$35y = 6{,}125$$
$$y = 175$$

name brand suit cost = $320
generic label suit cost = $175

65. let x = telephone sales
y = showroom sales

$$x + y = 40{,}000$$
$$0.05x + 0.06y = 2{,}250$$

$$-5(x + y) = (40{,}000)(-5)$$
$$100(0.05x + 0.06y = 2{,}250)$$

$$-5x - 5y = -200{,}000$$
$$5x + 6y = 225{,}000$$
$$\overline{}$$
$$y = 25{,}000 \text{ showroom sales}$$

$$x + y = 40{,}000$$
$$x + 25{,}000 = 40{,}000$$
$$x = 15{,}000 \text{ telephone sales}$$

telephone sales = $15,000
showroom sales = $25,000

Chapter 10 Practice Test

1.
$$2a + b = 10$$
$$a - b = 5$$

$$2a + b = 10$$
$$b = -2a + 10$$

$$m = \frac{-2}{1}$$
y-intercept = 10

$$a - b = 5$$
$$-b = -a + 5$$
$$\frac{-b}{-1} = \frac{-a}{-1} + \frac{5}{-1}$$
$$b = a - 5$$

$$m = \frac{1}{1}$$
y-intercept = -5 Solution: (5, 0)

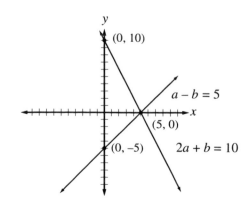

3. $3x + 4y = 6$
        $x + y = 5$

$3x + 4y = 6$ $x + y = 5$
  $4y = -3x + 6$ $y = -x + 5$
  $\dfrac{4y}{4} = \dfrac{-3x}{4} + \dfrac{6}{4}$ $m = -1;\ b = 5$
        Solution: $(14, -9)$
  $y = \dfrac{-3}{4}x + \dfrac{3}{2}$

$m = \dfrac{-3}{4};\ \ b = 1\dfrac{1}{2}$

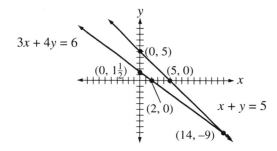

5. $2c - 3d = 6$
        $c - 12 = 3d$

$2c - 3d = 6$ $c - 12 = 3d$
  $-3d = -2c + 6$ $3d = c - 12$
  $\dfrac{-3d}{-3} = \dfrac{-2c}{-3} + \dfrac{6}{-3}$ $\dfrac{3d}{3} = \dfrac{c}{3} - \dfrac{12}{3}$
  $d = \dfrac{2}{3}c - 2$ $d = \dfrac{1}{3}c - 4$
  $m = \dfrac{2}{3};\ b = -2$ $m = \dfrac{1}{3};\ b = -4$
        Solution: $(-6, -6)$

7. $p + 2m = 0$
        $2p = -m$
        $-2(p + 2m) = (0)(-2)$
        $2p + m = 0$

        $-2p - 4m = 0$
        $\underline{2p + \ m = 0}$
        $-3m = 0$
        $m = 0$

        $p + 2m = 0$
        $p + 2(0) = 0$
        $p + 0 = 0$
        $p = 0$
        $p = 0,\ m = 0$
        Solution: $(0, 0)$

9. $3x + y = 5$
        $2x - y = 0$

        $3x + y = 5$
        $\underline{2x - y = 0}$
        $5x\quad\ \ = 5$
        $x = 1$

        $3x + y = 5$
        $3(1) + y = 5$
        $3 + y = 5$
        $y = 2$
        $x = 1,\ y = 2$
        Solution: $(1, 2)$

11. $4x + 3y = 14$
        $x - y = 0$
        $x = y$

        $4x + 3y = 14$
        $4(y) + 3y = 14$
        $4y + 3y = 14$
        $7y = 14$
        $y = 2$

        $x - y = 0$
        $x - 2 = 0$
        $x = 2$
        $x = 2,\ y = 2$
        Solution: $(2, 2)$

13. $7p + r = -6$
          $3p + r = 6$
          $7p + r = -6$
          $r = -6 - 7p$
          $3p + r = 6$
          $3p + (-6 - 7p) = 6$
          $-4p - 6 = 6$
          $-4p = 12$
          $p = -3$
          $7p + r = -6$
          $7(-3) + r = -6$
          $-21 + r = -6$
          $r = 15$
          $p = -3, r = 15$
          Solution: $(-3, 15)$

15. $38 + d = -3a$
          $5a + 1 = 4d$
          $4(3a + d) = (-38)(4)$
          $5a - 4d = -1$
          $12a + 4d = -152$
          $\underline{5a - 4d = -1}$
          $17a \qquad = -153$
          $a = -9$
          $38 + d = -3a$
          $38 + d = -3(-9)$
          $38 + d = 27$
          $d = -11$
          $a = -9, d = -11$
          Solution: $(-9, -11)$

17.   let x = current 1
          y = current 2
          $x + y = 35$
          $x - y = 5$
          $x + y = 35$
          $\underline{x - y = 5}$
          $2x \quad = 40$
          $x = 20$
          $x + y = 35$
          $20 + y = 35$
          $y = 15$
          larger current   = 20 A
          smaller current = 15 A

19.   let x = length
          y = width
          $x = 1.5y$
          $x - y = 17$
          $x - y = 17$
          $1.5y - y = 17$
          $0.5y = 17$
          $y = 34$
          $x = 1.5y$
          $x = 1.5(34)$
          $x = 51$
          length = 51 in.
          width = 34 in.

21.   let x = 3.5% investment
          y = 4.0% investment
          $x + y = 25,000$ .... Equation 1
          $0.035x + 0.04y = 900$ .... Equation 2
          $-35(x + y) = 25,000(-35)$ .... Equation 1
          $1,000(0.035x + 0.04y) = 900(1,000)$ .... Equation 2
          $-35x - 35y = -875,000$ .... Equation 1
          $35x + 40y = 900,000$ .... Equation 2
          $5y = 25,000$ .... Sum of Eq. 1 and 2
          $y = 5,000$
          $x + y = 25,000$
          $x + 5,000 = 25,000$
          $x = 25,000 - 5,000$
          $x = 20,000$
          $20,000 invested at 3.5%
          $5,000 invested at 4%

23.   let x = capacitance 1
          y = capacitance 2
          $x + y = 0.00027$
          $x - y = 0.00016$
          $x + y = 0.00027$
          $\underline{x - y = 0.00016}$
          $2x \quad = 0.00043$
          $x = 0.000215$
          $x + y = 0.00027$
          $0.000215 + y = 0.00027$
          $y = 0.000055$
          The two capacitances are 0.000215 F and 0.000055 F.

Chapters 7–10 Cumulative Practice Test

1. $4 - 3(2x - 5) =$
  $4 - 6x + 15 =$
  $-6x + 19$
  or
  $19 - 6x$

3. $7x - 3(x - 8) = 28$
  $7x - 3x + 24 = 28$
  $4x + 24 = 28$
  $4x = 4$
  $x = 1$

5. $\dfrac{5}{12}x - \dfrac{3}{4} = \dfrac{1}{9} - \dfrac{2}{3}x$

  $108\left(\dfrac{5}{12}x\right) - 108\left(\dfrac{3}{4}\right) = 108\left(\dfrac{1}{9}\right) - 108\left(\dfrac{2}{3}x\right)$

  $45x - 81 = 12 - 72x$
  $45x + 72x = 12 + 81$
  $117x = 93$
  $\dfrac{117x}{117} = \dfrac{93}{117}$
  $x = \dfrac{31}{39}$

7. $Z = \sqrt{R^2 + X^2}$
  $Z = \sqrt{4^2 + 7^2}$
  $Z = \sqrt{16 + 49}$
  $Z = \sqrt{65}$
  $Z = 8.1\ \Omega$

9. $\dfrac{2x}{7} = \dfrac{3}{5}$
  $2x(5) = 3(7)$
  $10x = 21$
  $x = \dfrac{21}{10}$

11. $\dfrac{400}{75} = \dfrac{x}{30}$ Let x = the number of teeth of smaller gear
  $75x = 400(30)$
  $75x = 12{,}000$
  $x = \dfrac{12{,}000}{75}$
  $x = 160$ teeth
There are 160 teeth in the small gear.

13. $y = -2x + 4$

x	y	$-2x + 4$
-2	8	$-2(-2) + 4 = 4 + 4$
0	4	$-2(0) + 4 = 0 + 4$
2	0	$-2(2) + 4 = -4 + 4$

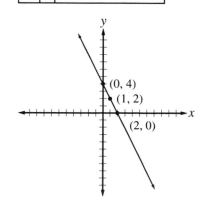

15. $y - \dfrac{3}{4}x = 2$

  $y = \dfrac{3}{4}x + 2$

  slope is $\dfrac{3}{4}$; y-intercept is $(0, 2)$

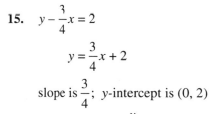
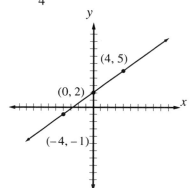

17. $3x - 2y = 10$

$-2y = -3x + 10$

$$\frac{-2y}{-2} = \frac{-3x}{-2} + \frac{10}{-2}$$

$$y = \frac{3}{2}x - 5$$

m (slope) $= \dfrac{3}{2}$;

b (y-intercept) $= 5$ or $(0, -5)$

19. $(5, -4)$ $m = -\dfrac{2}{3}$

$y - y_1 = m(x - x_1)$

$y - (-4) = -\dfrac{2}{3}(x - 5)$

$y + 4 = -\dfrac{2}{3}(x - 5)$

$y + 4 = -\dfrac{2}{3}x + \dfrac{2}{3}(5)$

$y = -\dfrac{2}{3}x + \dfrac{10}{3} + \dfrac{-12}{3}$

$y = -\dfrac{2}{3}x - \dfrac{2}{3}$

21. $(3, 2)$ and $(5, 6)$

$$m = \frac{y_2 - y_1}{x_2 - x_1}$$

$$m = \frac{6 - 2}{5 - 3}$$

$$m = \frac{4}{2}$$

$m = 2$

$y - y_1 = m(x - x_1)$

$y - 2 = 2(x - 3)$

$y - 2 = 2x - 6$

$y = 2x - 6 + 2$

$y = 2x - 4$

23. $m = 4$; $b = 3$

$y = mx + b$

$y = 4x + 3$

25. $x - y = 3$; $(5, 4)$

$-y = -x + 3$

$y = x - 3$ so $m = 1$

$y - y_1 = m(x - x_1)$

$y - 4 = 1(x - 5)$

$y - 4 = x - 5$

$-x + y = -5 + 4$

$-x + y = -1$

$x - y = 1$

27. $2x - y = -1$

$\underline{x + y = 4}$

$3x = 3$

$x = 1$

$x + y = 4$

$1 + y = 4$

$y = 3$

$x = 1, y = 3$

Solution: $(1, 3)$

29. Let $x =$ larger resistance

$y =$ smaller resistance

$x + y = 32$

$\underline{x - y = 8}$

$2x = 40$

$x = 20$

$x - y = 8$

$20 - y = 8$

$-y = 8 - 20$

$-y = -12$

$y = 12$

The larger resistance is 20 Ω. The smaller resistance is 12 Ω.

chapter **11** Powers and Polynomials

Chapter Review Exercises

1. $x^5 \cdot x^5 = x^{5+5} = x^{10}$

The exponent of the product is the *sum* of the exponents.

3. $3x^4 \cdot 7x^5 = 21x^9$

5. $\dfrac{x^8}{x^5} = x^{8-5} = x^3$

The exponent of the quotient is the *difference* between the exponents.

7. $\dfrac{21x^4}{3x} = 7x^{4-1} = 7x^3$

9. $\dfrac{x^3 y^{-1}}{x^2 y^2} = x^{3-2} y^{-1-2} = xy^{-3} = \dfrac{x}{y^3}$

11. $(x^3)^4 = x^{3(4)} = x^{12}$

To raise a power to a power, multiply exponents.

13. $(x^{-3})^{-5} = x^{-3(-5)} = x^{15}$

To raise a power to a power, multiply exponents.

15. $(-3x^2)^3 = (-3)^3(x^2)^3 = (-3)(-3)(-3)x^{2(3)} = -27x^6$

17. $\dfrac{xy^3}{xy^5} = x^{1-1} y^{3-5} = x^0 y^{-2} = \dfrac{1}{y^2}$

19. $5x^{-2} \cdot 8x^4 = 5(8)x^{-2+4} = 40x^2$

21. $7m^2 - 8m + 12m^4$

Degree of $12\,m^4$ is 4

Degree of polynomial: 4

23. $2x^3 y^2 - 15xy^3 + 21y^4$

Degree of $2x^3 y^2$ is $3 + 2 = 5$

Degree of polynomial: 5

25. $5x + 3x^3 - 8 + x^2$

Descending order: $3x^3 + x^2 + 5x - 8$
Degree: 3
Leading term: $3x^3$
Leading coefficient: 3

27. $5x^2 - 12x + 2x^4 - 32$

Descending order: $2x^4 + 5x^2 - 12x - 32$
Degree: 4
Leading term: $2x^4$
Leading coefficient: 2

29. $8x - 2x^4 - (3x^3 + 5x - x^3) = 8x - 2x^4 - 3x^3 - 5x + x^3 = -2x^4 - 2x^3 + 3x$

To simplify, remove grouping symbols, then combine like terms.

31. $4x^2 - (3y^2 + 7x^2 - 8y^2) = 4x^2 - 3y^2 - 7x^2 + 8y^2 = -3x^2 + 5y^2$

33. $-7x^8(-3x^{-2}) = (-7)(-3)(x^{8+(-2)}) = 21x^6$

Multiply the coefficients and add the exponents.

35. $2x(x^2 + 3x - 5) =$
$2x^3 + 6x^2 - 10x$

37. $-2x(x^3 - 7x^2 + 15) =$
$-2x^4 + 14x^3 - 30x$

39.
$$\overset{\text{F}\quad\text{L}}{(m + 3)(m - 7)} = \overset{\text{F}\quad\text{O}\quad\text{I}\quad\text{L}}{m^2 - 7m + 3m - 21}$$
$$= m^2 - 4m - 21$$

41.
$$\overset{\text{F}\quad\text{L}}{(4r - 5)(3r + 2)} = \overset{\text{F}\quad\text{O}\quad\text{I}\quad\text{L}}{12r^2 + 8r - 15r - 10}$$
$$= 12r^2 - 7r - 10$$

43.
$$(4 - 2m)(1 - 3m) = \overset{\text{F}\quad\text{O}\quad\text{I}\quad\text{L}}{4 - 12m - 2m + 6m^2}$$
$$= 4 - 14m + 6m^2$$

45.
$$(x + 3)(2x - 5) = \overset{\text{F}\quad\text{O}\quad\text{I}\quad\text{L}}{2x^2 - 5x + 6x - 15}$$
$$= 2x^2 + 1x - 15$$
$$= 2x^2 + x - 15$$

47.
$$(2a + 3b)(7a - b) = \overset{\text{F}\quad\text{O}\quad\text{I}\quad\text{L}}{14a^2 - 2ab + 21ab - 3b^2}$$
$$= 14a^2 + 19ab - 3b^2$$

49.
$$(9x - 2y)(3x + 4y) = \overset{\text{F}\quad\text{O}\quad\text{I}\quad\text{L}}{27x^2 + 36xy - 6xy - 8y^2}$$
$$= 27x^2 + 30xy - 8y^2$$

51.
$$(7m - 2n)(3m + 5n) = \overset{\text{F}\quad\text{O}\quad\text{I}\quad\text{L}}{21m^2 + 35mn - 6mn - 10n^2}$$
$$= 21m^2 + 29mn - 10n^2$$

53. $(5x - 3)(x^2 - 3x + 1) =$
$5x^3 - 15x^2 + 5x - 3x^2 + 9x - 3 =$
$5x^3 - 18x^2 + 14x - 3$

55. $(6x - 5)(6x + 5) = 36x^2 \quad - \quad 25$
square, minus, square

57. $(7y + 11)(7y - 11) = 49y^2 - 121$

59. $(8a - 5b)(8a + 5b) = 64a^2 \quad - \quad 25b^2$
square, minus, square

61. $\left(\sqrt{8} + 2\right)\left(\sqrt{8} - 2\right) = 8 - 4 = 4$

63. $(3 + i)(3 - i) = 9 - i^2 = 9 \quad - \quad (-1) \; = 10$
square, minus, square

65. $(x+9)^2 = x^2 \quad + \quad 18x \quad + \quad 81$
        square, double product, square

67. $(x-3)^2 = x^2 \quad - \quad 6x \quad + \quad 9$
        square, double product, square

69. $(4x-15)^2 = 16x^2 - \quad 120x \quad + 225$
        square, double product, square

71. $(8+7m)^2 = 64 \quad + \quad 112m \quad + 49m^2$
        square, double product, square

73. $(4x-11)^2 = 16x^2 - \quad 88x \quad + 121$
        square, double product, square

75. $(g-h)(g^2+gh+h^2) = g^3 \quad - \quad h^3$
        cube, minus, cube

77. $(2H-3T)(4H^2+6HT+9T^2) = 8H^3 \quad - \quad 27T^3$
        cube, minus, cube

79. $(6+i)(36-6i+i^2) = 216 \quad + \quad i^3$
        cube, plus, cube

81. $(z+2t)(z^2-2zt+4t^2) = z^3 \quad + \quad 8t^3$
        cube, plus, cube

83. $(7T+2)(49T^2-14T+4) = 343T^3 + \quad 8$
        cube, plus, cube

85. $\dfrac{12x^7}{-18x^4} = -\dfrac{2}{3}x^3 \text{ or } -\dfrac{2x^3}{3}$
Reduce coefficients and subtract the exponents.

87. $\dfrac{42x^3 y}{-15x^3 y^3} = -\dfrac{14}{5y^2}$

89. $\dfrac{6x^3 - 12x^2 + 21x}{3x} = \dfrac{6x^3}{3x} - \dfrac{12x^2}{3x} + \dfrac{21x}{3x}$
$= 2x^2 - 4x + 7$

91. $\dfrac{4x^5}{8x^2} - 3x^2(2x^4) =$
$\dfrac{1x^3}{2} - 6x^6 = \dfrac{x^3}{2} - 6x^6$

93. $\dfrac{16x^3 + 12x^4 - 20x^5}{-4x^2} = \dfrac{16x^3}{-4x^2} + \dfrac{12x^4}{-4x^2} - \dfrac{20x^5}{-4x^2} = -4x - 3x^2 + 5x^3$

95.
$$
\begin{array}{r}
x - 2 \\
x-9 \overline{\smash{\big)}\, x^2 - 11x + 18} \\
\underline{x^2 - 9x} \qquad \text{subtract} \\
-2x + 18 \\
\underline{-2x + 18} \quad \text{subtract} \\
0
\end{array}
$$

97.
$$
\begin{array}{r}
3x - 2 \\
5x+2 \overline{\smash{\big)}\, 15x^2 - 4x - 4} \\
\underline{15x^2 + 6x} \qquad \text{subtract} \\
-10x - 4 \\
\underline{-10x - 4} \quad \text{subtract} \\
0
\end{array}
$$

99.
$$
\begin{array}{r}
x^2 + x - 5 \\
x-5 \overline{\smash{\big)}\, x^3 - 4x^2 - 10x + 25} \\
\underline{x^3 - 5x^2} \qquad\qquad \text{subtract} \\
1x^2 - 10x \\
\underline{x^2 - 5x} \qquad \text{subtract} \\
-5x + 25 \\
\underline{-5x + 25} \quad \text{subtract} \\
0
\end{array}
$$

101.
$$
\begin{array}{r}
3x + 1 \\
2x^2 - x - 1 \overline{\smash{\big)}\, 6x^3 - x^2 - 4x - 1} \\
\underline{6x^3 - 3x^2 - 3x} \qquad \text{subtract} \\
2x^2 - x - 1 \\
\underline{2x^2 - x - 1} \quad \text{subtract} \\
0
\end{array}
$$

Chapter 11 Practice Test

1. $x^4(x) = x^4(x^1) = x^{4+1} = x^5$

3. $\left(\dfrac{4}{7}\right)^2 = \dfrac{4^2}{7^2} = \dfrac{16}{49}$

5. $\dfrac{x^{-7}}{x^3} = x^{-7-3} = x^{-10} = \dfrac{1}{x^{10}}$

7. $\left(\dfrac{x^2}{y}\right)^2 = \dfrac{(x^2)^2}{y^2} = \dfrac{x^4}{y^2}$

9. $4a(3a^2 - 2a + 5) = 12a^3 - 8a^2 + 20a$

11. $5x^2 - 3x - (2x + 4x^2) =$
$5x^2 - 3x - 2x - 4x^2 =$
$x^2 - 5x$

13. $-7x^3(-8x^4) = +56x^{3+4} = 56x^7$

15. $4xy - (3x - 2) + 4 = 4xy - 3x + 2 + 4 = 4xy - 3x + 6$

17. $14x - 3x + 21x$
Degree: 1

19. $6 - 3x^4 - 2x^3 = -3x^4 - 2x^3 + 6$

21. $(m - 7)(m + 7) = m^2 \quad - \quad 49$
square, minus, square

23. $(a + 3)^2 = a^2 \quad + \quad 6a \quad + \quad 9$
square, double product, square

25. $\overset{\text{F} \quad \text{O} \quad \text{I} \quad \text{L}}{(x - 3)(2x - 5)} = 2x^2 - 5x - 6x + 15$
$= 2x^2 - 11x + 15$

27. $(x - 2)(x^2 + 2x + 4) = x^3 \quad - \quad 8$
cube, minus, cube

29. $(5a - 3)(25a^2 + 15a + 9) = 125a^3 \quad - \quad 27$
cube, minus, cube

31.
$$\begin{array}{r}
2x + 7 \\
x - 3 \overline{\smash{\big)}\, 2x^2 + x - 21} \\
\underline{2x^2 - 6x} \qquad \text{subtract} \\
7x - 21 \\
\underline{7x - 21} \quad \text{subtract} \\
0
\end{array}$$

chapter 12   Roots and Radicals

Chapter Review Exercises

1. $\sqrt{49}$ radical notation
$49^{1/2}$ exponential notation

3. $\sqrt[4]{16}$ radical notation
$16^{1/4}$ exponential notation

5. $\sqrt{121}$ radical notation
$121^{1/2}$ exponential notation

7. $\sqrt{36}$ $\sqrt{38}$ $\sqrt{49}$
  ↓   ↓
  6   and   7
$\sqrt{38}$ is between 6 and 7.

9. $\sqrt{121}$ $\sqrt{135}$ $\sqrt{144}$
  ↓   ↓
 11   and   12
$\sqrt{135}$ is between 11 and 12.

11. $\sqrt[3]{27}$ $\sqrt[3]{60}$ $\sqrt[3]{64}$
  ↓   ↓
  3   and   4
$\sqrt[3]{60}$ is between 3 and 4.

13. $\sqrt{15} \approx 3.9$

15. $\sqrt{5} \approx 2.2$

17. $(x^7)^{1/2} = x^{7/2}$; $\sqrt{x^7}$

19. $\left(x^{1/3}\right)^2 = x^{2/3}$; $\left(\sqrt[3]{x}\right)^2$

21. $\sqrt{x} = x^{1/2}$

23. $\sqrt[5]{x^{4/5}} = x^{4/5}$

25. $\left(\sqrt[3]{xy}\right)^4 = \left((xy)^{1/3}\right)^4 = (xy)^{4/3} = x^{4/3}\,y^{4/3}$

27. $\sqrt{7} = 7^{1/2}$

29. $y^{3/5} = \sqrt[5]{y^3}$

31. $\sqrt{y^{12}} = (y^{12})^{1/2} = y^6$

33. $-\sqrt{b^{18}} = -(b^{18})^{1/2} = -b^9$

35. $\sqrt[3]{x} = x^{1/3}$

37. $\sqrt[5]{4y} = (4y)^{1/5}$ or $4^{1/5}\,y^{1/5}$

39. $\sqrt[3]{8b^{12}} = (8b^{12})^{1/3} = 8^{1/3}b^{12/3} = (2^3)^{1/3}b^4 = 2b^4$

41. $(a^{1/2})(a^{3/2}) = a^{(1/2)+(3/2)} = a^{4/2} = a^2$

43. $y^{3/4} \cdot y^{1/4} = y^{(3/4)+(1/4)} = y^{4/4} = y^1 = y$

45. $(3x^{1/4}y^2)^3 = (3)^3(x^{1/4})^3(y^2)^3 = 27x^{(1/4)(3)}y^{2(3)} = 27x^{3/4}y^6$

47. $(4ax^{1/2})^3 = (4)^3(a^1)^3(x^{1/2})^3 = 64a^{1(3)}x^{(1/2)(3)} = 64a^3x^{3/2}$

49. $\dfrac{x^{3/4}}{x^{1/4}} = x^{(3/4)-(1/4)} = x^{2/4} = x^{1/2}$

51. $\dfrac{a^{5/6}}{a^{-1/3}} = a^{(5/6)-(-1/3)} = a^{(5/6)+(2/6)} = a^{7/6}$

53. $\dfrac{x^{5/8}}{x^{3/4}} = x^{(5/8)-(3/4)} = x^{(5/8)-(6/8)} = x^{-1/8} = \dfrac{1}{x^{1/8}}$

55. $\dfrac{a^3}{a^{1/3}} = a^{3-(1/3)} = a^{(9/3)-(1/3)} = a^{8/3}$

57. $\dfrac{12a^4}{6a^{1/2}} = 2a^{4-(1/2)} = 2a^{(8/2)-(1/2)} = 2a^{7/2}$

$\dfrac{12(2)^4}{6(2^{1/2})} = \dfrac{192}{8.485281374} = 22.627417$

$2(2^{7/2}) = 22.627417$

59. $\dfrac{15a^{3/5}}{10a^5} = \dfrac{3}{2}a^{(3/5)-5} = \dfrac{3}{2}a^{(3/5)-(25/5)} = \dfrac{3}{2}a^{-22/5} = \dfrac{3}{2a^{22/5}}$

$\dfrac{15(2^{3/5})}{10(2^5)} = \dfrac{22.7357485}{320} = 0.0710492141$

$\dfrac{3}{2(2^{22/5})} = \dfrac{3}{42.22425314} = 0.0710492141$

61. $a^{2.3}(a^4) = a^{6.3};\qquad 2^{2.3}(2^4) = 78.79324245$

$\qquad\qquad\qquad\qquad\qquad\quad 2^{6.3} = 78.79324245$

63. $(4a^6b^8)^{1/2} = \sqrt{4a^6b^8} = \sqrt{4(64)(1)} = \sqrt{256} = 16$

$4^{1/2}(a^6)^{1/2}(b^8)^{1/2} = 2a^3b^4 = 2(2^3)(1)^4 = 16$

65. $\sqrt{x^2} = (x^2)^{1/2} = x$

67. $\sqrt{9P^3} = \sqrt{9}\,\sqrt{P^3} = \sqrt{9}\,\sqrt{P^2}\sqrt{P} = 3P\sqrt{P}$

69. $\sqrt{18a^2b} = \sqrt{18}\,\sqrt{a^2}\sqrt{b} = \sqrt{9\cdot 2}\,\sqrt{a^2}\sqrt{b} = (3\sqrt{2})(a)(\sqrt{b}) = 3a\sqrt{2b}$

71. $\sqrt{32x^5y^2} = \sqrt{32}\,\sqrt{x^5}\sqrt{y^2} = \sqrt{16\cdot 2}\sqrt{x^4\cdot x}\sqrt{y^2} = (4\sqrt{2})(x^2\sqrt{x})(y) = 4x^2y\sqrt{2x}$

73. $\sqrt{75x^{10}y^9} = \sqrt{75}\,\sqrt{x^{10}}\sqrt{y^9} = \sqrt{25\cdot 3}\,\sqrt{x^{10}}\sqrt{y^8\cdot y} = (5\sqrt{3})(x^5)(y^4\sqrt{y}) = 5x^5y^4\sqrt{3y}$

75. $\sqrt{147xy^8} = \sqrt{49\cdot 3xy^8} = 7y^4\sqrt{3x}$

77. $5\sqrt{3} - 7\sqrt{3} = -2\sqrt{3}$

79. $3\sqrt{7} - 2\sqrt{28} =$

$3\sqrt{7} - 2\sqrt{4\cdot 7} =$

$3\sqrt{7} - 2(2\sqrt{7}) =$

$3\sqrt{7} - 4\sqrt{7} =$

$-1\sqrt{7} =$

$-\sqrt{7}$

81. $2\sqrt{6} + 3\sqrt{54} =$

$2\sqrt{6} + 3\sqrt{9\cdot 6} =$

$2\sqrt{6} + 3(3\sqrt{6}) =$

$2\sqrt{6} + 9\sqrt{6} =$

$11\sqrt{6}$

83. $4\sqrt{3} - 8\sqrt{48} =$

$4\sqrt{3} - 8\sqrt{16\cdot 3} =$

$4\sqrt{3} - 8(4\sqrt{3}) =$

$4\sqrt{3} - 32\sqrt{3} =$

$-28\sqrt{3}$

85. $5\sqrt{8} - 3\sqrt{50} =$

$5\sqrt{4\cdot 2} - 3\sqrt{25\cdot 2} =$

$5(2\sqrt{2}) - 3(5\sqrt{2}) =$

$10\sqrt{2} - 15\sqrt{2} =$

$-5\sqrt{2}$

87. $3\sqrt{2} - 5\sqrt{32} =$

$3\sqrt{2} - 5\sqrt{16\cdot 2} =$

$3\sqrt{2} - 5(4\sqrt{2}) =$

$3\sqrt{2} - 20\sqrt{2} =$

$-17\sqrt{2}$

89. $2\sqrt{8}\cdot 3\sqrt{6} =$

$6\sqrt{48} =$

$6\sqrt{16\cdot 3} =$

$6(4\sqrt{3}) =$

$24\sqrt{3}$

91. $5\sqrt{3} \cdot 8\sqrt{7} = 40\sqrt{21}$

93. $-8\sqrt{5} \cdot 4\sqrt{30} = -32\sqrt{150}$
$$= -32\sqrt{25 \cdot 6}$$
$$= -32(5\sqrt{6})$$
$$= -160\sqrt{6}$$

95. $\sqrt{3}(\sqrt{12} - 5) =$
$$\sqrt{36} - 5\sqrt{3} =$$
$$6 - 5\sqrt{3}$$

97. $\sqrt{3}(\sqrt{6} - \sqrt{15}) =$
$$\sqrt{18} - \sqrt{45} =$$
$$\sqrt{9 \cdot 2} - \sqrt{9 \cdot 5} =$$
$$3\sqrt{2} - 3\sqrt{5}$$

99. $(\sqrt{5} - 8)(\sqrt{5} + 8) =$
$$\sqrt{25} + 8\sqrt{5} - 8\sqrt{5} - 64 =$$
$$5 - 64 =$$
$$-59$$

101. $\dfrac{3\sqrt{5}}{2\sqrt{20}} = \dfrac{3}{2\sqrt{4}} = \dfrac{3}{2(2)} = \dfrac{3}{4}$

103. $\dfrac{6\sqrt{18}}{8\sqrt{12}} = \dfrac{3\sqrt{3}}{4\sqrt{2}}$

or $\dfrac{3\sqrt{3}}{4\sqrt{2}} \cdot \dfrac{\sqrt{2}}{\sqrt{2}} = \dfrac{3\sqrt{6}}{4(2)} = \dfrac{3\sqrt{6}}{8}$

105. $\dfrac{5\sqrt{48}}{20\sqrt{20}} = \dfrac{\sqrt{12}}{4\sqrt{5}} = \dfrac{\sqrt{4 \cdot 3}}{4\sqrt{5}}$
$$= \dfrac{2\sqrt{3}}{4\sqrt{5}} = \dfrac{\sqrt{3}}{2\sqrt{5}}$$

or $\dfrac{\sqrt{3}}{2\sqrt{5}} \cdot \dfrac{\sqrt{5}}{\sqrt{5}} = \dfrac{\sqrt{15}}{10}$

107. $\dfrac{\sqrt{3y^3}}{\sqrt{y^3}} = \sqrt{3}$

109. $\left(\sqrt{\dfrac{9}{16}}\right)^2 = \left(\dfrac{3}{4}\right)^2 = \dfrac{9}{16}$

111. $\sqrt{\dfrac{36x^8}{81y^{10}}} = \dfrac{6x^4}{9y^5}$

113. $\dfrac{\sqrt{7}}{\sqrt{12}} = \dfrac{\sqrt{7}}{\sqrt{4 \cdot 3}} = \dfrac{\sqrt{7}}{2\sqrt{3}} = \dfrac{\sqrt{7}}{2\sqrt{3}} \cdot \dfrac{\sqrt{3}}{\sqrt{3}}$
$$= \dfrac{\sqrt{21}}{2(3)} = \dfrac{\sqrt{21}}{6}$$

115. $\dfrac{\sqrt{3}}{\sqrt{8}} = \dfrac{\sqrt{3}}{\sqrt{8}} \cdot \dfrac{\sqrt{2}}{\sqrt{2}} = \dfrac{\sqrt{6}}{\sqrt{16}} = \dfrac{\sqrt{6}}{4}$

or $\dfrac{\sqrt{3}}{\sqrt{8}} \cdot \dfrac{\sqrt{8}}{\sqrt{8}} = \dfrac{\sqrt{24}}{\sqrt{64}} = \dfrac{\sqrt{4 \cdot 6}}{8} = \dfrac{2\sqrt{6}}{8} = \dfrac{\sqrt{6}}{4}$

117. $\dfrac{5\sqrt{3}}{\sqrt{24}} = \dfrac{5\sqrt{3}}{\sqrt{3 \cdot 8}} = \dfrac{5\sqrt{3}}{\sqrt{3} \cdot 4 \cdot 2} = \dfrac{5 \cdot \sqrt{2}}{2\sqrt{2} \cdot \sqrt{2}} = \dfrac{5\sqrt{2}}{2\sqrt{4}} = \dfrac{5\sqrt{2}}{2 \cdot 2} = \dfrac{5\sqrt{2}}{4}$

119. $\dfrac{2\sqrt{5}}{\sqrt{18}} = \dfrac{2\sqrt{5}}{\sqrt{9 \cdot 2}} = \dfrac{2\sqrt{5}}{3\sqrt{2}} = \dfrac{2\sqrt{5}}{3\sqrt{2}} \cdot \dfrac{\sqrt{2}}{\sqrt{2}} = \dfrac{2\sqrt{10}}{3 \cdot 2} = \dfrac{2\sqrt{10}}{6} = \dfrac{\sqrt{10}}{3}$

121. $\sqrt{-100} = \sqrt{100(-1)} = 10i$

123. $\pm\sqrt{-24y^7} = \pm\sqrt{4 \cdot 6}\sqrt{-1}\sqrt{y^6 \cdot y}$
$$= \pm(2\sqrt{6})(i)(y^3\sqrt{y})$$
$$= \pm 2y^3 i\sqrt{6y}$$

125. $i^{14} = (i^4)^3 \cdot i^2$
$$= (1)^3 \cdot (-1)$$
$$= -1$$

127. $i^{77} = (i^4)^{19} \cdot i$
$$= 1 \cdot i$$
$$= i$$

129. $15i = 0 + 15i$

131. $-12i^5 = -12i^4 \cdot i$
$$= -12i$$
$$= 0 - 12i$$

133. $(5 + 3i) + (2 - 7i)$
$$5 + 3i + 2 - 7i$$
$$(5 + 2) + (3i - 7i)$$
$$7 - 4i$$

135. $\left(7 - \sqrt{-9}\right) + \left(4 + \sqrt{-16}\right)$
$(7 - 3i) + (4 + 4i)$
$7 - 3i + 4 + 4i$
$(7 + 4) + (-3i + 4i)$
$11 + i$

137. $(4i + 3)(4i - 3) =$
$16i^2 - 12i + 12i - 9 =$
$16(-1) - 9 =$
$-16 - 9 =$
-25

Chapter 12 Practice Test

1. $2\sqrt{7} \cdot 3\sqrt{2} = 6\sqrt{14}$

3. $4\sqrt{3} + 2\sqrt{3} = 6\sqrt{3}$

5. $\dfrac{4\sqrt{2}}{\sqrt{3}} = \dfrac{4\sqrt{2}}{\sqrt{3}} \cdot \dfrac{\sqrt{3}}{\sqrt{3}} = \dfrac{4\sqrt{6}}{\sqrt{9}} = \dfrac{4\sqrt{6}}{3}$
rationalize the denominator

7. $\dfrac{6\sqrt{8}}{2\sqrt{3}} = \dfrac{3\sqrt{8}}{\sqrt{3}} = \dfrac{3\sqrt{4 \cdot 2}}{\sqrt{3}} = \dfrac{3(2\sqrt{2})}{\sqrt{3}} = \dfrac{6\sqrt{2}}{\sqrt{3}}$
$= \dfrac{6\sqrt{2}}{\sqrt{3}} \cdot \dfrac{\sqrt{3}}{\sqrt{3}} = \dfrac{6\sqrt{6}}{\sqrt{9}} = \dfrac{6\sqrt{6}}{3} = 2\sqrt{6}$

9. $\dfrac{3\sqrt{5}}{2} \cdot \dfrac{7}{\sqrt{3x}} = \dfrac{21\sqrt{5}}{2\sqrt{3x}} = \dfrac{21\sqrt{5}}{2\sqrt{3x}} \cdot \dfrac{\sqrt{3x}}{\sqrt{3x}} = \dfrac{21\sqrt{15x}}{2\sqrt{9x^2}}$
$= \dfrac{21\sqrt{15x}}{2(3x)} = \dfrac{21\sqrt{15x}}{6x} = \dfrac{7\sqrt{15x}}{2x}$

11. $\sqrt[3]{27x^{15/3}} = 27^{1/3}x^{15/3}$
$= \left(3^3\right)^{1/3}x^5$
$= 3^1 x^5$
$= 3x^5$

13. $\sqrt[5]{x^{10}y^{15}z^{30}} = x^{10/5}y^{15/5}z^{30/5}$
$= x^2 y^3 z^6$

15. $\left(125x^{1/2}y^6\right)^{1/3} = 125^{1/3}\left(x^{1/2}\right)^{1/3}\left(y^6\right)^{1/3}$
$= 5x^{1/6}y^2$

17. $\dfrac{12x^{3/5}}{6x^{-2/5}} = 2x^{(3/5) - (-2/5)}$
$= 2x^{5/5} = 2x^1 = 2x$

19. $i^{23} = \left(i^4\right)^5 \cdot i^3$
$= 1(-i)$
$= -i$

21. $(5 + 3i) - (8 - 2i)$
$5 + 3i - 8 + 2i$
$5 - 8 + 3i + 2i$
$-3 + 5i$

23. $5\sqrt{3} + 8\sqrt{5} - 7\sqrt{3} =$
$-2\sqrt{3} + 8\sqrt{5}$

25. $\sqrt{7}(\sqrt{5} - 4) =$
$\sqrt{35} - 4\sqrt{7}$

27. $(5\sqrt{2} - 3)(5\sqrt{2} + 3) =$
$25\sqrt{4} + 15\sqrt{2} - 15\sqrt{2} - 9 =$
$25(2) - 9 =$
$50 - 9 =$
41

chapter 13   Factoring

Chapter Review Exercises

1. $5x + 5y = 5\left(\dfrac{5x}{5} + \dfrac{5y}{5}\right) = 5(x + y)$

3. $12m^2 - 8n^2 = 4\left(\dfrac{12m^2}{4} - \dfrac{8n^2}{4}\right)$
$= 4(3m^2 - 2n^2)$

5. $2a^3 - 14a^2 - 2a = 2a\left(\dfrac{2a^3}{2a} - \dfrac{14a^2}{2a} - \dfrac{2a}{2a}\right)$
$= 2a(a^2 - 7a - 1)$

7. $15x^3 - 5x^2 - 20x =$
$5x\left(\dfrac{15x^3}{5x} - \dfrac{5x^2}{5x} - \dfrac{20x}{5x}\right) =$
$5x(3x^2 - x - 4)$

9. $18a^3 - 12a^2 =$
$6a^2\left(\dfrac{18a^3}{6a^2} - \dfrac{12a^2}{6a^2}\right) =$
$6a^2(3a - 2)$

11. $-6x^2 - 10x =$
$-2x\left(\dfrac{-6x^2}{-2x} - \dfrac{10x}{-2x}\right) =$
$-2x(3x + 5)$

13. $5\sqrt{3} + 15\sqrt{7} =$
$5\left(\dfrac{5\sqrt{3}}{5} + \dfrac{15\sqrt{7}}{5}\right) =$
$5(\sqrt{3} + 3\sqrt{7})$

15. $\sqrt{12} - 10\sqrt{7} =$
$\sqrt{4 \cdot 3} - 10\sqrt{7} =$
$2\sqrt{3} - 10\sqrt{7} =$
$2\left(\dfrac{2\sqrt{3}}{2} - \dfrac{10\sqrt{7}}{2}\right) =$
$2(\sqrt{3} - 5\sqrt{7})$

17. $64 + 4a^2$ No; this is a *sum*, not a difference.

19. $H^2 - G^2$ Yes

21. $64b^2 - 49$ Yes

23. $-9x^2 + 12x + 4$ No; the first term is negative.

25. $9t^2 - 24tp + 16p^2$ Yes

27. $j^2 + 10j + 25$ Yes

29. $125a^3 - 8b^3$ Yes, difference

31. $8z^3 - 125$ Yes, difference

33. $64w^3 + 27$ Yes, sum

35. $25y^2 - 4 =$
$(5y + 2)(5y - 2)$

37. $a^2b^2 + 49$ NSP; this is a *sum*, not a difference.

39. $a^2 + 2a + 1 =$
$(a + 1)^2$
square root, middle sign, square root, quantity squared

41. $16c^2 - 24bc + 9b^2 =$
$(4c - 3b)^2$

43. $n^2 + 169 - 26n =$
$n^2 - 26n + 169 =$
$(n - 13)^2$

45. $36a^2 + 84ab + 49b^2 =$
$(6a + 7b)^2$

47. $49 - 14x + x^2 =$
$(7 - x)^2$

49. $64 + 25x^2$ NSP; this is a *sum*, not a difference.

51. $16x^2 + 24x + 9y^2$ NSP; missing y factor in middle term

53. $49 - 81y^2 =$
$(7 + 9y)(7 - 9y)$

55. $9x^2 - 100y^2 =$
$(3x + 10y)(3x - 10y)$

57. $9x^2 - 6xy + y^2 =$
$(3x - y)^2$

59. $9x^2y^2 - z^2 =$
$(3xy + z)(3xy - z)$

61. $x^2 + 4x + 4 =$
$(x + 2)^2$

63. $\dfrac{4}{25}x^2 - \dfrac{1}{16}y^2 =$
$\left(\dfrac{2}{5}x + \dfrac{1}{4}y\right)\left(\dfrac{2}{5}x - \dfrac{1}{4}y\right)$

65. $T^3 - 8 =$
$(T - 2)(T^2 + 2T + 4)$

67. $d^3 + 729 =$
$(d + 9)(d^2 - 9d + 81)$

69. $27k^3 + 64 =$
$(3k + 4)(9k^2 - 12k + 16)$

71. $x^2 + 11x + 24 =$ $\dfrac{24}{}$
$(x +)(x +)$ $1 \cdot 24$
$(x + 3)(x + 8)$ $2 \cdot 12$
$3 \cdot 8$ ★ $+3 + 8 = +11$
$4 \cdot 6$

73. $x^2 + 13x + 30 =$ $\dfrac{30}{}$
$(x +)(x +)$ $1 \cdot 30$
$(x + 3)(x + 10)$ $2 \cdot 15$
$3 \cdot 10$ ★ $+3 + 10 = +13$
$5 \cdot 6$

75. $x^2 - 9x + 8 =$ $\dfrac{8}{}$
$(x -)(x -)$ $1 \cdot 8$ ★ $-1 - 8 = -9$
$(x - 1)(x - 8)$ $2 \cdot 4$

77. $x^2 - 11x - 26 =$ $\dfrac{26}{}$
$(x +)(x -)$ $1 \cdot 26$
$(x + 2)(x - 13)$ $2 \cdot 13$ ★ $+2 - 13 = -11$

79. $x^2 + 5x - 24 =$ $\dfrac{24}{}$
$(x +)(x -)$ $1 \cdot 24$
$(x + 8)(x - 3)$ $2 \cdot 12$
$3 \cdot 8$ ★ $-3 + 8 = +5$
$4 \cdot 6$

81. $6x^2 + 25x + 4 =$ $6 \cdot 4 = 24$
$6x^2 + x + 24x + 4 =$ $1 \cdot 24$ ★ $+1 + 24 = +25$
$(6x^2 + x) + (24x + 4) =$ $2 \cdot 12$
$x(6x + 1) + 4(6x + 1) =$ $3 \cdot 8$
$(6x + 1)(x + 4)$ $4 \cdot 6$

83. $5x^2 - 34x + 24 =$ $5 \cdot 24 = 120$
$5x^2 - 4x - 30x + 24 =$ $1 \cdot 120$
$(5x^2 - 4x) + (-30x + 24) =$ $2 \cdot 60$
$3 \cdot 40$
$x(5x - 4) - 6(5x - 4) =$ $4 \cdot 30$ ★ $-4 - 30 = -34$
$(5x - 4)(x - 6)$ $5 \cdot 24$
$6 \cdot 20$
$8 \cdot 15$
$10 \cdot 12$

85. $6x^2 - x - 35 =$ $6 \cdot 35 = 210$
$6x^2 + 14x - 15x - 35 =$ $1 \cdot 210$
$2 \cdot 105$
$(6x^2 + 14x) + (-15x - 35) =$ $3 \cdot 70$
$2x(3x + 7) - 5(3x + 7) =$ $5 \cdot 42$
$(3x + 7)(2x - 5)$ $6 \cdot 35$
$7 \cdot 30$
$10 \cdot 21$
$14 \cdot 15$ ★ $+14 - 15 = -1$

87.
$$7x^2 - 13x - 24 =$$
$$7x^2 + 8x - 21x - 24 =$$
$$(7x^2 + 8x) + (-21x - 24) =$$
$$x(7x + 8) - 3(7x + 8) =$$
$$(7x + 8)(x - 3)$$

$7 \cdot 24 = 168$
$1 \cdot 168$
$2 \cdot 84$
$3 \cdot 56$
$4 \cdot 42$
$6 \cdot 28$
$7 \cdot 24$
$8 \cdot 21$ ★ $+8 - 21 = -13$
$12 \cdot 14$

89. $9a^2 - 100 =$
$(3a + 10)(3a - 10)$

91.
$$2x^2 - 3x - 2 =$$
$$2x^2 + x - 4x - 2 =$$
$$(2x^2 + x) + (-4x - 2) =$$
$$x(2x + 1) - 2(2x + 1) =$$
$$(2x + 1)(x - 2)$$

$2 \cdot 2 = 4$
$1 \cdot 4$ ★ $+1 - 4 = -3$
$2 \cdot 2$

93. $a^2 - 81 =$
$(a + 9)(a - 9)$

95. $y^2 - 14y + 49 =$
$(y - 7)^2$

97. $b^2 + 8b + 15 =$
$(b + \ )(b + \ )$
$(b + 3)(b + 5)$

15
$1 \cdot 15$
$3 \cdot 5$ ★ $+3 + 5 = +8$

99. $169 - m^2 =$
$(13 + m)(13 - m)$

101. $x^2 - 4x - 32 =$
$(x + \ )(x - \ )$
$(x + 4)(x - 8)$

32
$1 \cdot 32$
$2 \cdot 16$
$4 \cdot 8$ ★ $+4 - 8 = -4$

103. $x^2 + 19x - 20 =$
$(x + \ )(x - \ )$
$(x + 20)(x - 1)$

20
$1 \cdot 20$ ★ $-1 + 20 = +19$
$2 \cdot 10$
$4 \cdot 5$

105.
$$2x^2 - 4x - 16 =$$
$$2\left[\frac{2x^2}{2} - \frac{4x}{2} - \frac{16}{2}\right] =$$
$$2(x^2 - 2x - 8) =$$
$$2(x + \ )(x - \ )$$
$$2(x + 2)(x - 4)$$

8
$1 \cdot 8$
$2 \cdot 4$ ★ $+2 - 4 = -2$

107.
$$2x^3 - 10x^2 - 12x =$$
$$2x\left[\frac{2x^3}{2x} - \frac{10x^2}{2x} - \frac{12x}{2x}\right] =$$
$$2x(x^2 - 5x - 6) =$$
$$2x(x + \ )(x - \ )$$
$$2x(x + 1)(x - 6)$$

6
$1 \cdot 6$ ★ $+1 - 6 = -5$
$2 \cdot 3$

Chapter 13 Practice Test

1. $7x^2 + 8x = x\left(\dfrac{7x^2}{x} + \dfrac{8x}{x}\right)$

$\qquad\quad = x(7x + 8)$

3. $7a^2b - 14ab = 7ab\left(\dfrac{7a^2b}{7ab} - \dfrac{14ab}{7ab}\right)$

$\qquad\qquad\quad = 7ab(a - 2)$

5. $9x^2 - 25 = (3x + 5)(3x - 5)$

7. $x^2 - 18x + 81 = (x - 9)^2$

9. $27r^3 + 64s^3 = (3r + 4s)(9r^2 - 12rs + 16s^2)$

11.
$$6x^2 - 5x - 6 =$$
$$6x^2 + 4x - 9x - 6 =$$
$$(6x^2 + 4x) + (-9x - 6) =$$
$$2x(3x + 2) - 3(3x + 2) =$$
$$(3x + 2)(2x - 3)$$

$\begin{array}{l} 6 \cdot 6 = 36 \\ \hline 1 \cdot 36 \\ 2 \cdot 18 \\ 3 \cdot 12 \\ 4 \cdot 9 \quad \bigstar +4 - 9 = -5 \\ 6 \cdot 6 \end{array}$

13. $a^2 + 16ab + 64b^2 =$
$(a + 8b)^2$

15. $b^2 - 3b - 10$
$(b + \ )(b - \ )$
$(b + 2)(b - 5)$

$\begin{array}{l} 10 \\ \hline 1 \cdot 10 \\ 2 \cdot 5 \quad \bigstar +2 - 5 = -3 \end{array}$

17.
$$3m^2 - 5m + 2 =$$
$$3m^2 - 2m - 3m + 2 =$$
$$(3m^2 - 2m) + (-3m + 2) =$$
$$m(3m - 2) - 1(3m - 2) =$$
$$(3m - 2)(m - 1)$$

$\begin{array}{l} 3 \cdot 2 = 6 \\ \hline 1 \cdot 6 \\ 2 \cdot 3 \quad \bigstar -2 - 3 = -5 \end{array}$

19. $3x^2 - 12 =$
$3(x^2 - 4) =$
$3(x + 2)(x - 2)$

21. $5x^2 - 20 =$
$5(x^2 - 4) =$
$5(x + 2)(x - 2)$

23. $3x^2 + 12x + 12 =$
$3(x^2 + 4x + 4) =$
$3(x + 2)^2$

chapter 14

Rational Expressions and Equations

Chapter Review Exercises

1. $\dfrac{18}{24} = \dfrac{3 \cdot \cancel{6}}{4 \cdot \cancel{6}} = \dfrac{3}{4}$

3. $\dfrac{5a^2 b^3 c}{10a^3 bc^2} = \dfrac{5a^{2-3} b^{3-1} c^{1-2}}{10} = \dfrac{b^2}{2ac}$

5. $\dfrac{4xy(x-3)}{8xy(x+3)} = \dfrac{\cancel{4xy}(x-3)}{\underset{2}{\cancel{8xy}}(x+3)} = \dfrac{x-3}{2(x+3)}$

7. $\dfrac{(x-4)(x+2)}{(x+2)(4-x)} = \dfrac{(x-4)\cancel{(x+2)}}{\cancel{(x+2)}(4-x)} = \dfrac{x-4}{4-x} = -1$

9. $\dfrac{m^2 - n^2}{m^2 + n^2}$

11. $\dfrac{x}{x+xy} = \dfrac{x}{x(1+y)} = \dfrac{\cancel{x}}{\cancel{x}(1+y)} = \dfrac{1}{1+y}$

13. $\dfrac{5x+15}{x+3} = \dfrac{5(x+3)}{x+3} = \dfrac{5\cancel{(x+3)}}{\cancel{x+3}} = 5$

15. $\dfrac{y^2 + 2y + 1}{y+1} = \dfrac{(y+1)^2}{y+1} = \dfrac{\overset{y+1}{\cancel{(y+1)^2}}}{\cancel{y+1}} = y+1$

17. $\dfrac{2x-6}{x^2 + 3x - 18} = \dfrac{2(x-3)}{(x+6)(x-3)} = \dfrac{2\cancel{(x-3)}}{(x+6)\cancel{(x-3)}} = \dfrac{2}{x+6}$

19. $\dfrac{3x-9}{x-3} = \dfrac{3(x-3)}{x-3} = \dfrac{3\cancel{(x-3)}}{\cancel{x-3}} = 3$

21. $\dfrac{3x^2}{2y} \cdot \dfrac{5x}{6y} = \dfrac{\overset{1}{\cancel{3}}x^2}{2y} \cdot \dfrac{5x}{\underset{2}{\cancel{6}}y} = \dfrac{5x^3}{4y^2}$

23. $\dfrac{\overset{3}{\cancel{9}}}{x+b} \cdot \dfrac{5x+5b}{\cancel{3}} = \dfrac{3}{\cancel{x+b}} \cdot \dfrac{5\cancel{(x+b)}}{1} = 3(5) = 15$

25. $\dfrac{4y^2 - 4y + 1}{6y-6} \cdot \dfrac{24}{2y-1} = \dfrac{(2y-1)^2}{6(y-1)} \cdot \dfrac{24}{(2y-1)}$

$= \dfrac{\overset{2y-1}{\cancel{(2y-1)^2}}}{\underset{1}{\cancel{6}}(y-1)} \cdot \dfrac{\overset{4}{\cancel{24}}}{\cancel{2y-1}}$

$= \dfrac{4(2y-1)}{y-1}$

27. $\dfrac{5-x}{x-5} \cdot \dfrac{x-1}{1-x} = \dfrac{\overset{-1}{\cancel{5-x}}}{\cancel{x-5}} \cdot \dfrac{\overset{-1}{\cancel{x-1}}}{\cancel{1-x}}$

$= 1$

29.
$$\frac{x^2 + 6x + 9}{x^2 - 4} \cdot \frac{x - 2}{x + 3} = \frac{(x + 3)^2}{(x + 2)(x - 2)} \cdot \frac{(x - 2)}{(x + 3)}$$
$$= \frac{\cancel{(x+3)}^{2}\,^{x+3}}{(x + 2)\cancel{(x - 2)}} \cdot \frac{\cancel{(x - 2)}}{\cancel{(x + 3)}}$$
$$= \frac{x + 3}{x + 2}$$

31.
$$\frac{2a + b}{8} \div \frac{2a + b}{2} = \frac{2a + b}{8} \cdot \frac{2}{2a + b}$$
$$= \frac{\cancel{2a + b}}{\cancel{8}_{4}} \cdot \frac{\cancel{2}}{\cancel{2a + b}}$$
$$= \frac{1}{4}$$

33.
$$\frac{y^2 - 2y + 1}{y} \div \frac{1}{y - 1} = \frac{(y - 1)^2}{y} \cdot \frac{(y - 1)}{1}$$
$$= \frac{(y - 1)^3}{y}$$

35.
$$\frac{y^2 + 6y + 9}{y^2 + 4y + 4} \div \frac{y + 3}{y + 2} = \frac{y^2 + 6y + 9}{y^2 + 4y + 4} \cdot \frac{y + 2}{y + 3}$$
$$= \frac{(y + 3)^2}{(y + 2)^2} \cdot \frac{y + 2}{y + 3}$$
$$= \frac{\cancel{(y+3)}^{2}\,^{y+3}}{\cancel{(y+2)}^{2}\,_{y+2}} \cdot \frac{\cancel{y - 2}}{\cancel{y+3}}$$
$$= \frac{y + 3}{y + 2}$$

37.
$$\frac{3x^2 + 6x}{x} \div \frac{2x + 4}{x^2} = \frac{3x^2 + 6x}{x} \cdot \frac{x^2}{2x + 4}$$
$$= \frac{3x(x + 2)}{x} \cdot \frac{x^2}{2(x + 2)}$$
$$= \frac{3x\cancel{(x + 2)}}{\cancel{x}} \cdot \frac{x^2}{2\cancel{(x + 2)}}$$
$$= \frac{3x^2}{2}$$

39.
$$\frac{y^2 - 16}{y + 3} \div \frac{y - 4}{y^2 - 9} = \frac{(y + 4)(y - 4)}{(y + 3)} \cdot \frac{(y + 3)(y - 3)}{(y - 4)}$$
$$= \frac{(y + 4)\cancel{(y - 4)}}{\cancel{(y + 3)}} \cdot \frac{\cancel{(y + 3)}(y - 3)}{\cancel{(y - 4)}}$$
$$= (y + 4)(y - 3)$$

41.
$$\frac{\dfrac{5}{x - 3}}{4} = \frac{5}{x - 3} \cdot \frac{1}{4} = \frac{5}{4(x - 3)} = \frac{5}{4x - 12}$$

43.
$$\frac{\dfrac{x^2 - 4x}{6x}}{\dfrac{x - 4}{8x^2}} = \frac{x^2 - 4x}{6x} \cdot \frac{8x^2}{x - 4} = \frac{x(x - 4)}{6x} \cdot \frac{8x^2}{(x - 4)}$$
$$= \frac{x\cancel{(x - 4)}}{\cancel{6x}_{3}} \cdot \frac{\cancel{8}^{4}x^2}{\cancel{(x - 4)}}$$
$$= \frac{4x^2}{3}$$

45.
$$\frac{12}{6 - \sqrt{5}} = \frac{12}{(6 - \sqrt{5})} \cdot \frac{(6 + \sqrt{5})}{(6 + \sqrt{5})} = \frac{72 + 12\sqrt{5}}{36 - 5}$$
$$= \frac{72 + 12\sqrt{5}}{31}$$

47. $\dfrac{7+\sqrt{3}}{7-\sqrt{3}} = \dfrac{(7+\sqrt{3})}{(7-\sqrt{3})} \cdot \dfrac{(7+\sqrt{3})}{(7+\sqrt{3})} = \dfrac{49 + 7\sqrt{3} + 7\sqrt{3} + 3}{49 - 3}$

$$= \dfrac{52 + 14\sqrt{3}}{46} = \dfrac{2(26 + 7\sqrt{3})}{46}$$

$$= \dfrac{26 + 7\sqrt{3}}{23}$$

49. $\dfrac{5+\sqrt{2}}{7-3\sqrt{5}} = \dfrac{(5+\sqrt{2})}{(7-3\sqrt{5})} \cdot \dfrac{(7+3\sqrt{5})}{(7+3\sqrt{5})}$

$$= \dfrac{35 + 15\sqrt{5} + 7\sqrt{2} + 3\sqrt{10}}{49 - 45}$$

$$= \dfrac{35 + 15\sqrt{5} + 7\sqrt{2} + 3\sqrt{10}}{4}$$

51. $\dfrac{8+2\sqrt{3}}{5} = \dfrac{(8+2\sqrt{3})}{5} \cdot \dfrac{(8-2\sqrt{3})}{(8-2\sqrt{3})}$

$$= \dfrac{64 - 12}{40 - 10\sqrt{3}} = \dfrac{52}{2(20 - 5\sqrt{3})}$$

$$= \dfrac{26}{20 - 5\sqrt{3}}$$

53. $\dfrac{4-\sqrt{13}}{12} = \dfrac{(4-\sqrt{13})}{12} \cdot \dfrac{(4+\sqrt{13})}{(4+\sqrt{13})}$

$$= \dfrac{16 - 13}{48 + 12\sqrt{13}} = \dfrac{3}{3(16 + 4\sqrt{13})}$$

$$= \dfrac{1}{16 + 4\sqrt{13}}$$

55. $\dfrac{5-\sqrt{7}}{16} = \dfrac{(5-\sqrt{7})}{16} \cdot \dfrac{(5+\sqrt{7})}{(5+\sqrt{7})}$

$$= \dfrac{25 - 7}{16(5 + \sqrt{7})} = \dfrac{18}{16(5 + \sqrt{7})}$$

$$= \dfrac{9}{8(5 + \sqrt{7})} = \dfrac{9}{40 + 8\sqrt{7}}$$

57. $\dfrac{2}{9} + \dfrac{4}{9} = \dfrac{6}{9} = \dfrac{2 \cdot \cancel{3}}{3 \cdot \cancel{3}} = \dfrac{2}{3}$

59. $\dfrac{3x}{7} + \dfrac{2x}{14} = \dfrac{3x(2)}{7(2)} + \dfrac{2x}{14}$

$$= \dfrac{6x}{14} + \dfrac{2x}{14}$$

$$= \dfrac{8x}{14} = \dfrac{2(4x)}{2(7)}$$

$$= \dfrac{\cancel{2}(4x)}{\cancel{2}(7)}$$

$$= \dfrac{4x}{7}$$

61. $\dfrac{3x}{4} + \dfrac{5x}{6} = \dfrac{3x(3)}{4(3)} + \dfrac{5x(2)}{6(2)}$

$$= \dfrac{9x}{12} + \dfrac{10x}{12}$$

$$= \dfrac{19x}{12}$$

63. $\dfrac{5}{x} - \dfrac{7}{3} = \dfrac{(3)(5)}{(3)x} - \dfrac{7(x)}{3(x)}$

$$= \dfrac{15}{3x} - \dfrac{7x}{3x}$$

$$= \dfrac{15 - 7x}{3x}$$

65. $\dfrac{3}{4x} + \dfrac{2}{x} + \dfrac{3}{6x} = \dfrac{3(3)}{3(4x)} + \dfrac{12(2)}{12(x)} + \dfrac{2(3)}{2(6x)}$

$\qquad\qquad\quad = \dfrac{9}{12x} + \dfrac{24}{12x} + \dfrac{6}{12x}$

$\qquad\qquad\quad = \dfrac{39}{12x}$

$\qquad\qquad\quad = \dfrac{13}{4x}$

67. $\dfrac{7}{x-3} + \dfrac{3}{x+2} = \dfrac{7(x+2)}{(x-3)(x+2)} + \dfrac{3(x-3)}{(x-3)(x+2)}$

$\qquad\qquad\quad = \dfrac{7x + 14 + 3x - 9}{(x-3)(x+2)}$

$\qquad\qquad\quad = \dfrac{10x + 5}{(x-3)(x+2)}$

$\qquad\qquad\qquad\quad$ or

$\qquad\qquad\quad = \dfrac{5(2x+1)}{(x-3)(x+2)}$

69. $\dfrac{8}{x+3} - \dfrac{2}{x-4} = \dfrac{8(x-4)}{(x+3)(x-4)} - \dfrac{2(x+3)}{(x+3)(x-4)}$

$\qquad\qquad\quad = \dfrac{8x - 32 - 2x - 6}{(x+3)(x-4)}$

$\qquad\qquad\quad = \dfrac{6x - 38}{(x+3)(x-4)}$

$\qquad\qquad\qquad\quad$ or

$\qquad\qquad\quad = \dfrac{2(3x - 19)}{(x+3)(x-4)}$

71. $\dfrac{x}{x-5} - \dfrac{3}{5-x} = \dfrac{x}{x-5} - \dfrac{3(-1)}{x-5}$

$\qquad\qquad\quad = \dfrac{x + 3}{x - 5}$

73. $\dfrac{\dfrac{5}{x} - \dfrac{3}{4x}}{\dfrac{1}{3x} + \dfrac{2}{x}} = \dfrac{\dfrac{5(4)}{4x} - \dfrac{3}{4x}}{\dfrac{1}{3x} + \dfrac{2(3)}{3x}} =$

$\dfrac{\dfrac{20}{4x} - \dfrac{3}{4x}}{\dfrac{1}{3x} + \dfrac{6}{3x}} = \dfrac{\dfrac{17}{4x}}{\dfrac{7}{3x}}$

$\qquad\qquad = \dfrac{17}{4x} \cdot \dfrac{3x}{7} = \dfrac{17}{4\cancel{x}} \cdot \dfrac{3\cancel{x}}{7}$

$\qquad\qquad = \dfrac{51}{28}$

75. $\dfrac{\dfrac{3x}{6} - \dfrac{5}{x}}{\dfrac{x}{3} + \dfrac{4}{2x}} = \dfrac{\dfrac{3x(x)}{6x} - \dfrac{5(6)}{6x}}{\dfrac{x(2x)}{6x} + \dfrac{4(3)}{6x}}$

$\qquad\qquad = \dfrac{\dfrac{3x^2 - 30}{6x}}{\dfrac{2x^2 + 12}{6x}}$

$\qquad\qquad = \dfrac{3x^2 - 30}{6x} \cdot \dfrac{6x}{2x^2 + 12}$

$\qquad\qquad = \dfrac{3(x^2 - 10)}{\cancel{6x}} \cdot \dfrac{\cancel{6x}}{2(x^2 + 6)}$

$\qquad\qquad = \dfrac{3(x^2 - 10)}{2(x^2 + 6)}$

$\qquad\qquad = \dfrac{3x^2 - 30}{2x^2 + 12}$

77. $\dfrac{4}{x} = \dfrac{3}{x-2}$

$\quad x = 0 \quad x - 2 = 0$

$\qquad\qquad\qquad x = 2$

Excluded values: 0, 2

79. $\dfrac{5}{2x-1} - \dfrac{6}{x} = \dfrac{4}{3x}$

$\qquad 2x - 1 = 0$

$\qquad\quad 2x = 1 \qquad 3x = 0$

$\qquad\quad \dfrac{2x}{2} = \dfrac{1}{2} \qquad \dfrac{3x}{3} = \dfrac{0}{3}$

$\qquad\quad x = \dfrac{1}{2} \qquad\quad x = 0$

Excluded values: 0, $\dfrac{1}{2}$

81. $\dfrac{4}{x} = \dfrac{1}{x+5}$

$4(x+5) = 1(x)$ Excluded values: $0, -5$

$4x + 20 = x$

$20 = x - 4x$

$20 = -3x$

$\dfrac{20}{-3} = \dfrac{-3x}{-3}$

$x = -\dfrac{20}{3}$

83. $\dfrac{-4x}{x+1} = 3 - \dfrac{4}{x+1}$

$(x+1)\dfrac{-4x}{(x+1)} = (x+1)3 - (x+1)\dfrac{4}{(x+1)}$ Excluded value: -1

$(x+1)\dfrac{-4x}{(x+1)} = (x+1)3 - (x+1)\dfrac{4}{(x+1)}$

$-4x = 3x + 3 - 4$

$-4x = 3x - 1$

$-4x - 3x = -1$

$-7x = -1$

$\dfrac{-7x}{-7} = \dfrac{-1}{-7}$

$x = \dfrac{1}{7}$

85. $x =$ students in original group

$\dfrac{120}{x} - 10 = \dfrac{120}{4}$

$120(4) - 40x = 120x$

$480 = 160x$

$x = 3$ original students

87. $\dfrac{1}{5}(3) + \dfrac{1}{x}(3) = 1$

$\dfrac{3}{5} + \dfrac{3}{x} = 1$

$5x\left(\dfrac{3}{5}\right) + 5x\left(\dfrac{3}{x}\right) = 5x(1)$

$3x + 15 = 5x$

$15 = 5x - 3x$

$15 = 2x$

$\dfrac{15}{2} = \dfrac{2x}{2}$

$x = 7\dfrac{1}{2}$ hours alone

Chapter 14 Practice Test

1. $\dfrac{x-3}{2x-6} = \dfrac{x-3}{2(x-3)} = \dfrac{\cancel{x-3}}{2(\cancel{x-3})} = \dfrac{1}{2}$

3. $\dfrac{6x^2 - 11x + 4}{2x^2 + 5x - 3} = \dfrac{\dfrac{(6x - \ )(6x - \ )}{6}}{\dfrac{(2x + \ )(2x - \ )}{2}}$ $\begin{aligned} 6 \cdot 4 &= 24 \\ \overline{3 \cdot 8} \quad &-3 - 8 = -11 \\ 2 \cdot 3 &= \ 6 \\ \overline{1 \cdot 6} \quad &-1 + 6 = 5 \end{aligned}$

$\qquad = \dfrac{\dfrac{(6x - 3)(6x - 8)}{6}}{\dfrac{(2x + 6)(2x - 1)}{2}}$

$\qquad = \dfrac{\dfrac{(\cancel{3})(2x - 1)(\cancel{2})(3x - 4)}{\cancel{6}}}{\dfrac{(\cancel{2})(x + 3)(2x - 1)}{\cancel{2}}}$

$\qquad = \dfrac{(2x - 1)(3x - 4)}{(x + 3)(2x - 1)} = \dfrac{(\cancel{2x - 1})(3x - 4)}{(x + 3)(\cancel{2x - 1})}$

$\qquad = \dfrac{3x - 4}{x + 3}$

5. $\dfrac{(x-2)(x-4)}{(4-x)(x+2)} = \dfrac{(x-2)(\cancel{x-4})}{-(\cancel{x-4})(x+2)} = -\dfrac{x-2}{x+2} \ \text{or} \ \dfrac{2-x}{x+2}$

7. $\dfrac{6xy}{ab} \cdot \dfrac{a^2 b}{2xy^2} = \dfrac{6xya^2 b}{2xy^2 ab} = 3x^{1-1}y^{1-2}a^{2-1}b^{1-1} = \dfrac{3a}{y}$

9. $\dfrac{x - 2y}{x^3 - 3x^2 y} \div \dfrac{x^2 - 4y^2}{x - 3y} = \dfrac{x - 2y}{x^3 - 3x^2 y} \cdot \dfrac{x - 3y}{x^2 - 4y^2}$

$\qquad = \dfrac{(x - 2y)}{x^2(x - 3y)} \cdot \dfrac{(x - 3y)}{(x + 2y)(x - 2y)}$

$\qquad = \dfrac{(\cancel{x - 2y})}{x^2(\cancel{x - 3y})} \cdot \dfrac{(\cancel{x - 3y})}{(x + 2y)(\cancel{x - 2y})}$

$\qquad = \dfrac{1}{x^2(x + 2y)}$

11. $\dfrac{2x^2 + 3x + 1}{x} \div \dfrac{x + 1}{1} = \dfrac{2x^2 + 3x + 1}{x} \cdot \dfrac{1}{x + 1}$

$\qquad = \dfrac{(2x + 1)(x + 1)}{x} \cdot \dfrac{1}{x + 1}$

$\qquad = \dfrac{(2x + 1)(\cancel{x + 1})}{x} \cdot \dfrac{1}{(\cancel{x + 1})}$

$\qquad = \dfrac{2x + 1}{x}$

13. $\dfrac{1}{x + 2} - \dfrac{1}{x - 3} = \dfrac{1(x - 3)}{(x + 2)(x - 3)} - \dfrac{1(x + 2)}{(x + 2)(x - 3)}$

$\qquad = \dfrac{x - 3 - x - 2}{(x + 2)(x - 3)}$

$\qquad = \dfrac{-5}{(x + 2)(x - 3)} \ \text{or} \ -\dfrac{5}{(x + 2)(x - 3)}$

15. $\dfrac{3}{x} + \dfrac{1}{4} = \dfrac{3(4)}{4x} + \dfrac{1(x)}{4x}$

$\qquad = \dfrac{12 + x}{4x}$

17.

$$\frac{5}{3x-2} + \frac{7}{2-3x} = \frac{5}{3x-2} + \frac{7}{-3x+2}$$

$$= \frac{5}{3x-2} + \frac{7}{-1(3x-2)}$$

$$= \frac{5}{3x-2} - \frac{7}{3x-2}$$

$$= -\frac{2}{3x-2} \text{ or } \frac{2}{2-3x}$$

19.

$$\frac{x-2y}{x^2-4y^2} = \frac{x-2y}{(x+2y)(x-2y)}$$

$$= \frac{(x-2y)}{(x+2y)(x-2y)}$$

$$= \frac{1}{x+2y}$$

21.

$$\frac{2x}{1-\frac{3}{x}} = \frac{2x}{\frac{x}{x}-\frac{3}{x}} = \frac{2x}{\frac{x-3}{x}}$$

$$= \frac{2x}{1} \cdot \frac{x}{x-3} = \frac{2x^2}{x-3}$$

23.

$$\frac{5}{x} = \frac{2}{x+3}$$

$$x = 0 \quad x+3 = 0$$

$$x = -3$$

Excluded values: 0, −3

25.

$$\frac{3x}{x-2} + 4 = \frac{3}{x-2} \quad \text{Excluded value: 2}$$

$$(x-2)\frac{3x}{(x-2)} + (x-2)4 = (x-2)\frac{3}{(x-2)}$$

$$3x + 4(x-2) = 3$$

$$3x + 4x - 8 = 3$$

$$7x - 8 = 3$$

$$7x = 3 + 8$$

$$7x = 11$$

$$\frac{7x}{7} = \frac{11}{7}$$

$$x = \frac{11}{7}$$

Check:

$$\frac{3\left(\frac{11}{7}\right)}{\frac{11}{7}-2} + 4 = \frac{3}{\frac{11}{7}-2}$$

$$\frac{\frac{33}{7}}{\frac{11}{7}-\frac{14}{7}} + 4 = \frac{3}{\frac{11}{7}-\frac{14}{7}}$$

$$\frac{\frac{33}{7}}{-\frac{3}{7}} + 4 = \frac{3}{-\frac{3}{7}}$$

$$\frac{33}{7}\left(-\frac{7}{3}\right) + 4 = \frac{3}{1}\left(-\frac{7}{3}\right)$$

$$-11 + 4 = -7$$

$$-7 = -7$$

27.

$$\frac{5}{x-2} = \frac{-4}{x+1} \quad \text{Excluded values: 2, −1}$$

$$5(x+1) = -4(x-2)$$

$$5x + 5 = -4x + 8$$

$$5x + 4x = 8 - 5$$

$$9x = 3$$

$$\frac{9x}{9} = \frac{3}{9}$$

$$x = \frac{3}{9}$$

$$x = \frac{1}{3}$$

29.

x = number of people in first group
$2x$ = number of people in second group

$$\frac{100,000}{x} - 10,000 = \frac{100,000}{2x}$$

$$2(100,000) - 2x(10,000) = 100,000$$

$$200,000 - 20,000x = 100,000$$

$$-20,000x = -100,000$$

$$x = 5 \text{ people in original group}$$

Quadratic and Higher-Degree Equations

Chapter Review Exercises

1. $x^2 = 49$
$x^2 - 49 = 0$
$a = 1; c = -49$
pure

3. $5x^2 - 45 = 0$
$a = 5; c = -45$
pure

5. $8x^2 + 6x = 0$
$a = 8; b = 6$
incomplete

7. $x^2 - 32 = 0$
$a = 1; c = -32$
pure

9. $3x^2 + 6x + 1 = 0$
$a = 3; b = 6; c = 1$
complete

11. $2x^2 - 5 = 8x$
$2x^2 - 8x - 5 = 0$
standard form

13. $5 + x^2 - 7x = 0$
$x^2 - 7x + 5 = 0$
standard form

15. $3x = 1 - 4x^2$
$4x^2 + 3x - 1 = 0$
standard form

17. $x^2 = 100$
$x = \pm\sqrt{100}$
$x = \pm 10$
$x = 10$ or $x = -10$

19. $4x^2 = 9$
$\dfrac{4x^2}{4} = \dfrac{9}{4}$
$x^2 = \dfrac{9}{4}$
$x = \pm\sqrt{\dfrac{9}{4}}$
$x = \pm\dfrac{\sqrt{9}}{\sqrt{4}}$
$x = \pm\dfrac{3}{2}$ or ± 1.5
$x = 1.5$ or $x = -1.5$

21. $0.36y^2 = 1.09$
$\dfrac{0.36y^2}{0.36} = \dfrac{1.09}{0.36}$
$y^2 = 3.027777778$
$y \approx \pm\sqrt{3.027777778}$
$y \approx \pm 1.740$
$y \approx 1.740$ or $y \approx -1.740$

23. $5x^2 = 40$
$\dfrac{5x^2}{5} = \dfrac{40}{5}$
$x^2 = 8$
$x = \pm\sqrt{8}$
$x = \pm 2\sqrt{2}$
or
$x \approx \pm 2.828$
$x \approx 2.828$ or $x \approx -2.828$

25. $6x^2 + 4 = 34$
$6x^2 = 34 - 4$
$6x^2 = 30$
$\dfrac{6x^2}{6} = \dfrac{30}{6}$
$x^2 = 5$
$x = \pm\sqrt{5}$
or
$x \approx \pm 2.236$
$x \approx 2.236$ or $x \approx -2.236$

27. $3x^2 = 12$
$\dfrac{3x^2}{3} = \dfrac{12}{3}$
$x^2 = 4$
$x = \pm\sqrt{4}$
$x = \pm 2$
$x = 2$ or $x = -2$

29. $2x^2 = 34$
$\dfrac{2x^2}{2} = \dfrac{34}{2}$
$x^2 = 17$
$x = \pm\sqrt{17}$
or
$x \approx \pm 4.123$
$x \approx 4.123$ or $x \approx -4.123$

31.
$$3y^2 - 36 = -8$$
$$3y^2 = -8 + 36$$
$$3y^2 = 28$$
$$\frac{3y^2}{3} = \frac{28}{3}$$
$$y^2 = 9.333333333$$
$$y = \pm\sqrt{9.333333333}$$
$$y \approx \pm 3.055$$
$$y \approx 3.055 \text{ or } y \approx -3.055$$

33.
$$\frac{1}{2}x^2 = 8$$
$$(2)\frac{1}{2}x^2 = (2)8$$
$$x^2 = 16$$
$$x = \pm\sqrt{16}$$
$$x = \pm 4$$
$$x = 4 \text{ or } x = -4$$

35.
$$\frac{1}{4}x^2 - 1 = 15$$
$$\frac{1}{4}x^2 = 15 + 1$$
$$\frac{1}{4}x^2 = 16$$
$$(4)\frac{1}{4}x^2 = (4)16$$
$$x^2 = 64$$
$$x = \pm\sqrt{64}$$
$$x = \pm 8$$
$$x = 8 \text{ or } x = -8$$

37.
$$\pi r^2 = A$$
$$\pi(r^2) = 845$$
$$r^2 = \frac{845}{\pi}$$
$$r = \pm\sqrt{268.9718538}$$
$$r = \pm 16.4 \text{ cm}$$
$$r = 16.4 \text{ cm}$$
(negative root not appropriate)

39.
$$x^2 - 5x = 0$$
$$x(x - 5) = 0$$
$$x = 0 \text{ or } x - 5 = 0$$
$$x = 5$$
$$x = 0 \text{ or } x = 5$$

41.
$$6x^2 - 12x = 0$$
$$6x(x - 2) = 0$$
$$6x = 0 \text{ or } x - 2 = 0$$
$$\frac{6x}{6} = \frac{0}{6} \quad x = 2$$
$$x = 0$$
$$x = 0 \text{ or } x = 2$$

43.
$$10x^2 + 5x = 0$$
$$5x(2x + 1) = 0$$
$$5x = 0 \text{ or } 2x + 1 = 0$$
$$\frac{5x}{5} = \frac{0}{5} \qquad 2x = -1$$
$$x = 0 \qquad \frac{2x}{2} = \frac{-1}{2}$$
$$x = \frac{-1}{2} \text{ or } -\frac{1}{2}$$
$$x = 0 \text{ or } x = -\frac{1}{2}$$

45.
$$y^2 - 7y = 0$$
$$y(y - 7) = 0$$
$$y = 0 \text{ or } y - 7 = 0$$
$$y = 7$$
$$y = 0 \text{ or } y = 7$$

47.
$$12x^2 + 8x = 0$$
$$4x(3x + 2) = 0$$
$$4x = 0 \text{ or } 3x + 2 = 0$$
$$\frac{4x}{4} = \frac{0}{4} \qquad 3x = -2$$
$$x = 0 \qquad \frac{3x}{3} = \frac{-2}{3}$$
$$x = \frac{-2}{3} = -\frac{2}{3}$$
$$x = 0 \text{ or } x = -\frac{2}{3}$$

49.
$$x^2 + 3x = 0$$
$$x(x + 3) = 0$$
$$x = 0 \text{ or } x + 3 = 0$$
$$x = -3$$
$$x = 0 \text{ or } x = -3$$

51.
$$5x^2 = 45x$$
$$5x^2 - 45x = 0$$
$$5x(x - 9) = 0$$
$$5x = 0 \text{ or } x - 9 = 0$$
$$\frac{5x}{5} = \frac{0}{5} \qquad x = 9$$
$$x = 0$$
$$x = 0 \text{ or } x = 9$$

53.
$$y^2 + 8y = 0$$
$$y(y + 8) = 0$$
$$y = 0 \text{ or } y + 8 = 0$$
$$y = -8$$
$$y = 0 \text{ or } y = -8$$

55. $3m^2 - 5m = 0$

$m(3m - 5) = 0$

$m = 0$ or $3m - 5 = 0$

$3m = 5$

$\dfrac{3m}{3} = \dfrac{5}{3}$

$m = \dfrac{5}{3}$

$m = 0$ or $m = \dfrac{5}{3}$

57. $2x^2 = x$

$2x^2 - x = 0$

$x(2x - 1) = 0$

$x = 0$ or $2x - 1 = 0$

$2x = 1$

$\dfrac{2x}{2} = \dfrac{1}{2}$

$x = \dfrac{1}{2}$

$x = 0$ or $x = \dfrac{1}{2}$

59. $3x^2 = 12x$

$3x^2 - 12x = 0$

$3x(x - 4) = 0$

$3x = 0$ or $x - 4 = 0$

$\dfrac{3x}{3} = \dfrac{0}{3}$ $x = 4$

$x = 0$

$x = 0$ or $x = 4$

61. $x^2 - 4x + 3 = 0$

$(x - 3)(x - 1) = 0$

$x - 3 = 0$ or $x - 1 = 0$

$x = 3$ or $x = 1$

63. $x^2 + 3x = 10$

$x^2 + 3x - 10 = 0$

$(x + 5)(x - 2) = 0$

$x + 5 = 0$ or $x - 2 = 0$

$x = -5$ or $x = 2$

65. $x^2 + 7x = -6$

$x^2 + 7x + 6 = 0$

$(x + 1)(x + 6) = 0$

$x + 1 = 0$ or $x + 6 = 0$

$x = -1$ or $x = -6$

67. $x^2 - 6x + 8 = 0$

$(x - 2)(x - 4) = 0$

$x - 2 = 0$ or $x - 4 = 0$

$x = 2$ or $x = 4$

69. $6y^2 - 5y - 6 = 0$

$6y^2 - 9y + 4y - 6 = 0$ $6 \cdot 6 = \underline{36}$

$(6y^2 - 9y) + (4y - 6) = 0$ $1 \cdot 36$

$3y(2y - 3) + 2(2y - 3) = 0$ $2 \cdot 18$

$(2y - 3)(3y + 2) = 0$ $3 \cdot 12$

$2y - 3 = 0$ or $3y + 2 = 0$ $4 \cdot 9$ $\blacktriangle$ ($(4 - 9 = -5)$)

$2y = 3$ $3y = -2$ $6 \cdot 6$

$\dfrac{2y}{2} = \dfrac{3}{2}$ $\dfrac{3y}{3} = \dfrac{-2}{3}$

$y = \dfrac{3}{2}$ or $y = -\dfrac{2}{3}$

71. $10y^2 - 21y - 10 = 0$

$10y^2 - 25y + 4y - 10 = 0$ $10 \cdot 10 = \underline{100}$

$(10y^2 - 25y) + (4y - 10) = 0$ $1 \cdot 100$

$5y(2y - 5) + 2(2y - 5) = 0$ $2 \cdot 50$

$(2y - 5)(5y + 2) = 0$ $4 \cdot 25$ $\bigstar$ ($+4 - 25 = -21$)

$2y - 5 = 0$ $5y + 2 = 0$ $5 \cdot 20$

$2y = 5$ $5y = -2$ $10 \cdot 10$

$\dfrac{2y}{2} = \dfrac{5}{2}$ $\dfrac{5y}{5} = \dfrac{-2}{5}$

$y = \dfrac{5}{2}$ or $y = -\dfrac{2}{5}$

73.
$$4x^2 + 7x + 3 = 0$$
$$4x^2 + 3x + 4x + 3 = 0 \qquad 4 \cdot 3 = \underline{\quad 12 \quad}$$
$$(4x^2 + 3x) + (4x + 3) = 0 \qquad \overline{1 \cdot 12}$$
$$x(4x + 3) + 1(4x + 3) = 0 \qquad 2 \cdot 6$$
$$(4x + 3)(x + 1) = 0 \qquad 3 \cdot 4 \quad \bigstar (3 + 4 = 7)$$
$$4x + 3 = 0 \text{ or } x + 1 = 0$$
$$4x = -3 \qquad x = -1$$
$$\frac{4x}{4} = \frac{-3}{4}$$
$$x = -\frac{3}{4}$$
$$x = -\frac{3}{4} \text{ or } x = -1$$

75.
$$12y^2 - 5y - 3 = 0$$
$$12y^2 + 4y - 9y - 3 = 0 \qquad 12 \cdot 3 = 36$$
$$(12y^2 + 4y) + (-9y - 3) = 0 \qquad \overline{1 \cdot 36}$$
$$4y(3y + 1) - 3(3y + 1) = 0 \qquad 2 \cdot 18$$
$$(3y + 1)(4y - 3) = 0 \qquad 3 \cdot 12$$
$$3y + 1 = 0 \text{ or } 4y - 3 = 0 \qquad 4 \cdot 9 \quad \bigstar (+4 - 9 = -5)$$
$$3y = -1 \qquad 4y = 3 \qquad 6 \cdot 6$$
$$\frac{3y}{3} = \frac{-1}{3} \qquad \frac{4y}{4} = \frac{3}{4}$$
$$y = -\frac{1}{3} \quad \text{ or } \quad y = \frac{3}{4}$$

77.
$$x^2 + 19x = 42$$
$$x^2 + 19x - 42 = 0 \qquad \underline{\quad 42 \quad}$$
$$(x - 2)(x + 21) = 0 \qquad \overline{1 \cdot 42}$$
$$x - 2 = 0 \text{ or } x + 21 = 0 \qquad 2 \cdot 21 \quad \bigstar (-2 + 21 = 19)$$
$$x = 2 \quad \text{ or } \quad x = -21 \qquad 3 \cdot 14$$
$$6 \cdot 7$$

79.
$$3y^2 + y - 2 = 0$$
$$3y^2 - 2y + 3y - 2 = 0 \qquad 3 \cdot 2 = \underline{\quad 6 \quad}$$
$$(3y^2 - 2y) + (3y - 2) = 0 \qquad \overline{1 \cdot 6}$$
$$y(3y - 2) + 1(3y - 2) = 0 \qquad 2 \cdot 3 \quad \bigstar (-2 + 3 = 1)$$
$$(3y - 2)(y + 1) = 0$$
$$3y - 2 = 0 \text{ or } y + 1 = 0$$
$$3y = 2 \qquad y = -1$$
$$\frac{3y}{3} = \frac{2}{3}$$
$$y = \frac{2}{3}$$
$$y = \frac{2}{3} \text{ or } y = -1$$

81.
$$2x^2 - 10x + 12 = 0$$
$$2(x^2 - 5x + 6) = 0$$
$$2(x - 2)(x - 3) = 0$$
$$x - 2 = 0 \text{ or } x - 3 = 0$$
$$x = 2 \text{ or } x = 3$$

83.
$$x^2 - 3x - 18 = 0$$
$$(x - 6)(x + 3) = 0$$
$$x - 6 = 0 \text{ or } x + 3 = 0$$
$$x = 6 \quad \text{or} \quad x = -3$$

85.
$$2y^2 + 22y + 60 = 0$$
$$2(y^2 + 11y + 30) = 0$$
$$2(y + 5)(y + 6) = 0$$
$$y + 5 = 0 \text{ or } y + 6 = 0$$
$$y = -5 \text{ or } y = -6$$

87.
$$x^2 + 7x - 18 = 0$$
$$(x + 9)(x - 2) = 0$$
$$x + 9 = 0 \text{ or } x - 2 = 0$$
$$x = -9 \quad \text{or} \quad x = 2$$

89.
$$A = lw; \, l = w + 7$$
$$A = (w + 7)w$$
$$228 = w^2 + 7w$$
$$0 = w^2 + 7w - 228$$
$$0 = (w + 19)(w - 12)$$
$$w + 19 = 0 \quad \text{or} \qquad w - 12 = 0$$
$$w = -19 \qquad\qquad w = 12 \text{ ft}$$
Disregard negative root.
$$l = w + 7$$
$$l = 12 + 7$$
$$l = 19 \text{ ft}$$

91.
$$x^2 - 4x + 4 = 0$$
$$x^2 - 4x = -4$$
$$x^2 - 4x + \left(\frac{4}{2}\right)^2 = -4 + \left(\frac{4}{2}\right)^2$$
$$x^2 - 4x + 2^2 = -4 + 2^2$$
$$x^2 - 4x + 4 = -4 + 4$$
$$x^2 - 4x + 4 = 0$$
$$(x - 2)^2 = 0$$
$$x - 2 = 0 \text{ or } x - 2 = 0$$
$$x = 2 \text{ or } \quad x = 2 \text{ double root}$$

93.
$$x^2 - 8x + 12 = 0$$
$$x^2 - 8x = -12$$
$$x^2 - 8x + \left(\frac{8}{2}\right)^2 = -12 + \left(\frac{8}{2}\right)^2$$
$$x^2 - 8x + (4)^2 = -12 + (4)^2$$
$$x^2 - 8x + 16 = -12 + 16$$
$$(x - 4)^2 = 4$$
$$x - 4 = \pm 2$$
$$x - 4 = 2 \text{ or } x - 4 = -2$$
$$x = 6 \text{ or } \quad x = 2$$

95.
$$x^2 - 8x + 14 = 0$$
$$x^2 - 8x = -14$$
$$x^2 - 8x + \left(\frac{8}{2}\right)^2 = -14 + \left(\frac{8}{2}\right)^2$$
$$x^2 - 8x + (4)^2 = -14 + (4)^2$$
$$x^2 - 8x + 16 = -14 + 16$$
$$(x - 4)^2 = 2$$
$$x - 4 = \pm \sqrt{2}$$
$$x = 4 + \sqrt{2} \text{ or } x = 4 - \sqrt{2}$$

97.
$$x^2 - 6x + 12 = 0$$
$$x^2 - 6x = -12$$
$$x^2 - 6x + \left(\frac{6}{2}\right)^2 = -12 + \left(\frac{6}{2}\right)^2$$
$$x^2 - 6x + (3)^2 = -12 + (3)^2$$
$$x^2 - 6x + 9 = -12 + 9$$
$$(x - 3)^2 = -3$$
$$x - 3 = \pm \sqrt{-3}$$
$$x = 3 \pm i\sqrt{3}$$
$$x = 3 + i\sqrt{3} \text{ or } x = 3 - i\sqrt{3}$$
or no real roots

99.
$$x^2 - 5x + 4 = 0$$
$$x^2 - 5x = -4$$
$$x^2 - 5x + \left(\frac{5}{2}\right)^2 = -4 + \left(\frac{5}{2}\right)^2$$
$$x^2 - 5x + \frac{25}{4} = -4 + \frac{25}{4}$$
$$\left(x - \frac{5}{2}\right)^2 = \frac{-16}{4} + \frac{25}{4}$$
$$\left(x - \frac{5}{2}\right)^2 = \frac{9}{4}$$
$$x - \frac{5}{2} = \pm \frac{3}{2}$$
$$x = \frac{5}{2} + \frac{3}{2} \quad \text{or} \quad x = \frac{5}{2} - \frac{3}{2}$$
$$x = \frac{8}{2} = 4 \quad \text{or} \quad x = \frac{2}{2} = 1$$

101. $x^2 - 3x - 7 = 0$

$x^2 - 3x = 7$

$x^2 - 3x + \left(\dfrac{3}{2}\right)^2 = 7 + \left(\dfrac{3}{2}\right)^2$

$x^2 - 3x + \dfrac{9}{4} = 7 + \dfrac{9}{4}$

$\left(x - \dfrac{3}{2}\right)^2 = \dfrac{28}{4} + \dfrac{9}{4}$

$\left(x - \dfrac{3}{2}\right)^2 = \dfrac{37}{4}$

$x - \dfrac{3}{2} = \pm\sqrt{\dfrac{37}{4}}$

$x = \dfrac{3}{2} \pm \dfrac{\sqrt{37}}{2}$

$x = \dfrac{3 + \sqrt{37}}{2}$ or $x = \dfrac{3 - \sqrt{37}}{2}$

103. $x^2 - 2x = 8$

$x^2 - 2x - 8 = 0$

$a = 1;\ b = -2;\ c = -8$

105. $x^2 + 3x = 4$

$x^2 + 3x - 4 = 0$

$a = 1;\ b = 3;\ c = -4$

107. $x^2 - 3x = -2$

$x^2 - 3x + 2 = 0$

$a = 1;\ b = -3;\ c = 2$

109. $x^2 - 8x - 9 = 0$

$a = 1;\ b = -8;\ c = -9$

$x = \dfrac{-b \pm\sqrt{b^2 - 4ac}}{2a}$

$x = \dfrac{-(-8) \pm\sqrt{(-8)^2 - 4(1)(-9)}}{2(1)}$

$x = \dfrac{8 \pm\sqrt{64 + 36}}{2}$

$x = \dfrac{8 \pm\sqrt{100}}{2} = \dfrac{8 \pm 10}{2}$

$x = \dfrac{8 + 10}{2} = \dfrac{18}{2} = 9$

$x = \dfrac{8 - 10}{2} = \dfrac{-2}{2} = -1$

$x = 9$ or $x = -1$

111. $x^2 + 2x = 8$

$x^2 + 2x - 8 = 0$

$a = 1;\ b = 2;\ c = -8$

$x = \dfrac{-b \pm\sqrt{b^2 - 4ac}}{2a}$

$x = \dfrac{-2 \pm\sqrt{(2)^2 - 4(1)(-8)}}{2(1)}$

$x = \dfrac{-2 \pm\sqrt{4 + 32}}{2}$

$x = \dfrac{-2 \pm\sqrt{36}}{2} = \dfrac{-2 \pm 6}{2}$

$x = \dfrac{-2 + 6}{2} = \dfrac{4}{2} = 2$

$x = \dfrac{-2 - 6}{2} = \dfrac{-8}{2} = -4$

$x = 2$ or $x = -4$

113. $2x^2 - 3x - 2 = 0$

$a = 2;\ b = -3;\ c = -2$

$x = \dfrac{-b \pm\sqrt{b^2 - 4ac}}{2a}$

$x = \dfrac{-(-3) \pm\sqrt{(-3)^2 - 4(2)(-2)}}{2(2)}$

$x = \dfrac{3 \pm\sqrt{9 + 16}}{4}$

$x = \dfrac{3 \pm\sqrt{25}}{4}$

$x = \dfrac{3 \pm 5}{4}$

$x = \dfrac{3 + 5}{4} = \dfrac{8}{4} = 2$

$x = \dfrac{3 - 5}{4} = \dfrac{-2}{4} = \dfrac{-1}{2}$

$x = 2$ or $x = -\dfrac{1}{2}$

115. $2x^2 - 3x - 1 = 0$

$a = 2; b = -3; c = -1$

$$x = \frac{-b \pm \sqrt{b^2 - 4ac}}{2a}$$

$$x = \frac{-(-3) \pm \sqrt{(-3)^2 - 4(2)(-1)}}{2(2)}$$

$$x = \frac{3 \pm \sqrt{9 + 8}}{4}$$

$$x = \frac{3 \pm \sqrt{17}}{4}$$

$$x = \frac{3 \pm 4.123105626}{4}$$

$$x \approx \frac{3 + 4.123105626}{4} \approx \frac{7.123105626}{4} \approx 1.78$$

$$x \approx \frac{3 - 4.123105626}{4} \approx \frac{-1.123105626}{4} \approx -0.28$$

$x \approx 1.78$ or $x \approx -0.28$

117. $3x^2 + 5x + 1 = 0$

$a = 3; b = 5; c = 1$

$$x = \frac{-b \pm \sqrt{b^2 - 4ac}}{2a}$$

$$x = \frac{-5 \pm \sqrt{5^2 - 4(3)(1)}}{2(3)}$$

$$x = \frac{-5 \pm \sqrt{25 - 12}}{6}$$

$$x = \frac{-5 \pm \sqrt{13}}{6}$$

$$x \approx \frac{-5 \pm 3.605551275}{6}$$

$$x \approx \frac{-5 + 3.605551275}{6} \approx \frac{-1.394448725}{6} \approx -0.23$$

$$x \approx \frac{-5 - 3.605551275}{6} \approx \frac{-8.605551275}{6} \approx -1.43$$

$x \approx -0.23$ or $x \approx -1.43$

119.

$l = 2w$

$A = lw$

$A = (2w)(w)$

$A = 2w^2$

$240 = 2w^2$

$0 = 2w^2 - 240$

$a = 2; \ b = 0; \ c = -240$

$$w = \frac{-b \pm \sqrt{b^2 - 4ac}}{2a}$$

$$w = \frac{-(0) \pm \sqrt{(0)^2 - 4(2)(-240)}}{2(2)}$$

$$w = \frac{0 \pm \sqrt{0 + 1,920}}{4} = \frac{0 \pm \sqrt{1,920}}{4}$$

$$w \approx \frac{0 \pm 43.8178046}{4} \approx \frac{0 + 43.8178046}{4} \approx \frac{43.8178046}{4} \approx 11 \text{ ft}$$

$$w \approx \frac{0 - 43.8178046}{4} \approx \frac{-43.8178046}{4} \approx -11 \text{ ft} \quad \text{disregard negative solution}$$

$w \approx 11$ ft (rounded)

$l = 2w \approx 2(11) \approx 22$ ft

121. $l = 3w$

$A = lw$

$A = (3w)(w)$

$591 = 3w^2$

$0 = 3w^2 - 591$

$a = 3;\ b = 0;\ c = -591$

$$w = \frac{-b \pm \sqrt{b^2 - 4ac}}{2a} = \frac{-0 \pm \sqrt{0^2 - 4(3)(-591)}}{2(3)}$$

$$w = \frac{0 \pm \sqrt{0 + 7{,}092}}{6} \approx \frac{0 \pm 84.21401309}{6}$$

$$w = \frac{0 + 84.21401309}{6} \approx \frac{84.21401309}{6} \approx 14$$

$$w = \frac{0 - 84.21401309}{6} \approx \frac{-84.21401309}{6} \approx -14 \text{ disregard negative solution}$$

$w \approx 14$ in.

$l = 3w \approx 3(14) \approx 42$ in.

123. $y = -x^2 - 1$

x-coordinate of the vertex: $\dfrac{-b}{2a} = \dfrac{-(0)}{2(-1)} = \dfrac{0}{-2} = 0$

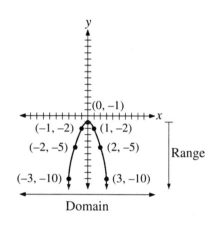

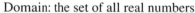
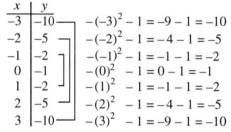

x	y	
−3	−10	$-(-3)^2 - 1 = -9 - 1 = -10$
−2	−5	$-(-2)^2 - 1 = -4 - 1 = -5$
−1	−2	$-(-1)^2 - 1 = -1 - 1 = -2$
0	−1	$-(0)^2 - 1 = 0 - 1 = -1$
1	−2	$-(1)^2 - 1 = -1 - 1 = -2$
2	−5	$-(2)^2 - 1 = -4 - 1 = -5$
3	−10	$-(3)^2 - 1 = -9 - 1 = -10$

Domain: the set of all real numbers

Range: the set of real numbers less than or equal to −1

125. $y = x^2 - 6x + 8$

x-coordinate of the vertex: $\dfrac{-b}{2a} = \dfrac{-(-6)}{2(1)} = \dfrac{6}{2} = 3$

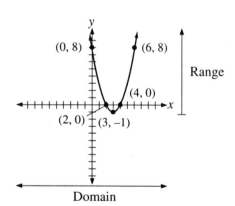

x	y	
0	8	$0^2 - 6(0) + 8 = 8$
1	3	$1^2 - 6(1) + 8 = 3$
2	0	$2^2 - 6(2) + 8 = 0$
3	−1	$3^2 - 6(3) + 8 = -1$
4	0	$4^2 - 6(4) + 8 = 0$
5	3	$5^2 - 6(5) + 8 = 3$
6	8	$6^2 - 6(6) + 8 = 8$

Domain: the set of all real numbers

Range: the set of real numbers greater than or equal to −1

127. $y = -x^2 + 2x - 1$

x-coordinate of the vertex: $\dfrac{-b}{2a} = \dfrac{-(2)}{2(-1)} = \dfrac{-2}{-2} = 1$

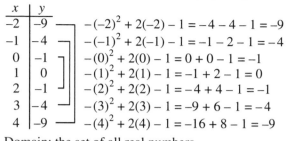

x	y	
-2	-9	$-(-2)^2 + 2(-2) - 1 = -4 - 4 - 1 = -9$
-1	-4	$-(-1)^2 + 2(-1) - 1 = -1 - 2 - 1 = -4$
0	-1	$-(0)^2 + 2(0) - 1 = 0 + 0 - 1 = -1$
1	0	$-(1)^2 + 2(1) - 1 = -1 + 2 - 1 = 0$
2	-1	$-(2)^2 + 2(2) - 1 = -4 + 4 - 1 = -1$
3	-4	$-(3)^2 + 2(3) - 1 = -9 + 6 - 1 = -4$
4	-9	$-(4)^2 + 2(4) - 1 = -16 + 8 - 1 = -9$

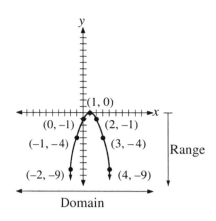

Domain: the set of all real numbers

Range: the set of real numbers less than or equal to 0

129. $y = x^2 - 2x - 8$

Axis of symmetry: $x = -\dfrac{b}{2a}$

$x = -\dfrac{-2}{2(1)}$

$x = 1$

Vertex: $(1, y)$

$y = x^2 - 2x - 8$

$y = 1^2 - 2(1) - 8$

$y = 1 - 2 - 8$

$y = -9$

Vertex: $(1, -9)$

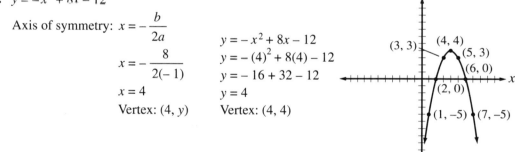

131. $y = -x^2 + 8x - 12$

Axis of symmetry: $x = -\dfrac{b}{2a}$

$x = -\dfrac{8}{2(-1)}$

$x = 4$

Vertex: $(4, y)$

$y = -x^2 + 8x - 12$

$y = -(4)^2 + 8(4) - 12$

$y = -16 + 32 - 12$

$y = 4$

Vertex: $(4, 4)$

133. $x^2 - 4x - 12 = 0$

$f(x) = x^2 - 4x - 12$

x	y	
★ −2	0	x-intercept or solution of equation
0	−12	
2	−16	vertex
4	−12	
★ 6	0	x-intercept or solution of equation

Two solutions: $x = -2$ or $x = 6$

Calculator:

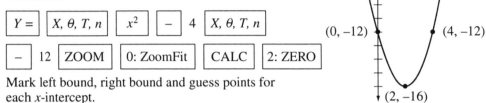

Mark left bound, right bound and guess points for each x-intercept.

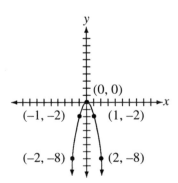

135. $x^2 + 8x = -16$

$f(x) = x^2 + 8x + 16$

x	y	
−2	4	$(-2)^2 + 8(-2) + 16 = 4 - 16 + 16$
★ −4	0	$(-4)^2 + 8(-4) + 16 = 16 - 32 + 16$
−6	4	$(-6)^2 + 8(-6) + 16 = 36 - 48 + 16$

One solution: $x = -4$ double root

Calculator:

Mark left bound, right bound and guess points for the x-intercept

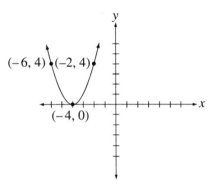

137. $y = -2x^2$

x-coordinate of the vertex: $\dfrac{-b}{2a} = \dfrac{-(0)}{2(-2)} = \dfrac{0}{-4} = 0$

x	y
−2	−8
−1	−2
0	0
1	−2
2	−8

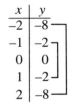

Calculator:

139. $x^2 + 8x + 16 = 0$

$a = 1; b = 8; c = 16$

$b^2 - 4ac = 8^2 - 4(1)(16)$

$\qquad = 64 - 64$

$\qquad = 0$

2 equal, real, rational, roots (double root)

141. $5x^2 - 100 = 0$

$a = 5; b = 0; c = -100$

$b^2 - 4ac = 0^2 - 4(5)(-100)$

$\qquad = 0 + 2,000$

$\qquad = 2,000$

2 unequal, real, irrational roots

143. $2x = 5x^2 - 3$

$0 = 5x^2 - 2x - 3$

$a = 5; b = -2; c = -3$

$b^2 - 4ac = (-2)^2 - 4(5)(-3)$

$\qquad = 4 + 60$

$\qquad = 64$ perfect square

2 unequal, real, rational roots

145. $3x - 2x^3 + 8 = 0$

$0 = 2x^3 - 3x - 8$

3rd degree or cubic equation

147. $6 - 3x - 3 = 2x + 4$

$0 = 2x + 3x + 4 + 3 - 6$

$0 = 5x + 1$

1st degree or linear equation

149. $5y^8 + 2y^3 - 6 = y^2$

$5y^8 + 2y^3 - y^2 - 6 = 0$

8th degree

151. $2x(3x - 2)(x - 2) = 0$

$\quad 2x = 0$ or $3x - 2 = 0$ or $x - 2 = 0$

$\quad \dfrac{2x}{2} = \dfrac{0}{2} \qquad 3x = 2 \qquad\qquad x = 2$

$\qquad x = 0 \qquad \dfrac{3x}{3} = \dfrac{2}{3}$

$\qquad\qquad\qquad\quad x = \dfrac{2}{3}$

$\qquad x = 0$ or $x = \dfrac{2}{3}$ or $x = 2$

153. $2x^3 + 10x^2 + 12x = 0$

$2x(x^2 + 5x + 6) = 0$

$2x(x + 2)(x + 3) = 0$

$\quad 2x = 0$ or $x + 2 = 0$ or $x + 3 = 0$

$\quad \dfrac{2x}{2} = \dfrac{0}{2} \qquad x = -2 \qquad x = -3$

$\qquad x = 0$

$\qquad x = 0$ or $x = -2$ or $x = -3$

155.

$\qquad 2x^3 + 9x^2 = 5x$

$\qquad 2x^3 + 9x^2 - 5x = 0$

$\qquad x(2x^2 + 9x - 5) = 0$

Factor trinomial:

$\qquad 2x^2 + 9x - 5$

$\qquad 2x^2 - x + 10x - 5$

$\qquad (2x^2 - x) + (10x - 5)$

$\qquad x(2x - 1) + 5(2x - 1)$

$\qquad\qquad x(2x - 1)(x + 5) = 0$ Use all three factors.

$\qquad\quad x = 0$ or $2x - 1 = 0$ or $x + 5 = 0$

$\qquad\qquad\qquad\quad 2x = 1 \qquad\quad x = -5$

$\qquad\qquad\qquad\quad \dfrac{2x}{2} = \dfrac{1}{2}$

$\qquad\qquad\qquad\qquad x = \dfrac{1}{2}$

$\qquad\quad x = 0$ or $x = \dfrac{1}{2}$ or $x = -5$

$2 \cdot 5 = \underline{\quad 10 \quad}$

$\qquad\quad 1 \cdot 10 \; \star (-1 + 10 = 9)$

$\qquad\quad 2 \cdot 5$

157. $3x^3 - 6x^2 = 0$

$3x^2(x - 2) = 0$ $x - 2 = 0$

$3x^2 = 0$

$\dfrac{3x^2}{3} = \dfrac{0}{3}$ $x = 2$

$x^2 = 0$

$\sqrt{x^2} = \sqrt{0}$

$x = 0$ or $x = 2$ (double root)

159. $x^3 + 6x^2 + 8x = 0$

$x(x^2 + 6x + 8) = 0$

$x(x + 2)(x + 4) = 0$

$x = 0$ or $x + 2 = 0$ or $x + 4 = 0$

$x = -2$ $x = -4$

$x = 0$ or $x = -2$ or $x = -4$

161. $x^3 - x^2 - 20x = 0$

$x(x^2 - x - 20) = 0$

$x(x + 4)(x - 5) = 0$

$x = 0$ or $x + 4 = 0$ or $x - 5 = 0$

$x = -4$ $x = 5$

$x = 0,\ x = -4,\ x = 5$

163. $y^3 - 6y^2 + 7y = 0$

$y(y^2 - 6y + 7) = 0$

$a = 1;\ b = -6;\ c = 7$

$y = \dfrac{-b \pm \sqrt{b^2 - 4ac}}{2a}$

$y = 0$ $y = \dfrac{-(-6) \pm \sqrt{36 - 4(1)(7)}}{2(1)}$

$y = \dfrac{6 \pm \sqrt{8}}{2}$

$y = \dfrac{6 \pm 2\sqrt{2}}{2}$

$y = \dfrac{\cancel{2}(3 \pm \sqrt{2})}{\cancel{2}}$

$y = 3 \pm \sqrt{2}$ exact roots

$y \approx 3 \pm 1.414$

$y \approx 4.414$ or $y \approx 1.586$ approximate roots

165. $x^3 - 3x^2 - 4x = 0$

$x(x^2 - 3x - 4) = 0$

$x(x + 1)(x - 4) = 0$

$x = 0$ $x + 1 = 0$ $x - 4 = 0$

$x = -1$ $x = 4$

167. $2y^3 + 6y^2 + 4y = 0$

$2y(y^2 + 3y + 2) = 0$

$2y(y + 2)(y + 1) = 0$

$2y = 0$ $y + 2 = 0$ $y + 1 = 0$

$y = 0$ $y = -2$ $y = -1$

169. $y = 5x^3$

x	y	$5x^3$
-2	-40	$5(-2)^3 = 5(-8) = -40$
-1	-5	$5(-1)^3 = 5(-1) = -5$
0	0	$5(0)^3\ \ = 5(0) = 0$
1	5	$5(1)^3\ \ = 5(1) = 5$
2	40	$5(2)^3\ \ = 5(8) = 40$

Domain: the set of all real numbers

Range: the set of all real numbers

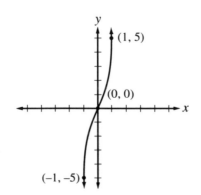

171. $x = -y^3 + 2y^2 + 2y + 3$

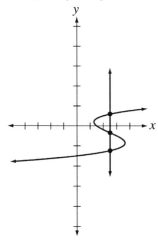

Not a function. Vertical line intersects curve in more than one point.

173. $\dfrac{x^2}{25} + \dfrac{y^2}{4} = 1$

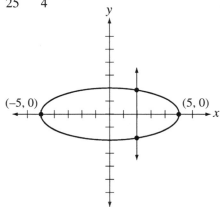

Not a function. Vertical line intersects two points of the graph

175.

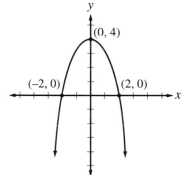

Domain: the set of all real numbers
Range: the set of real numbers less than or equal to 4

177.

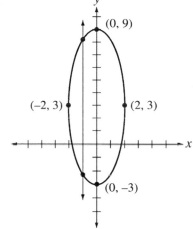

Domain: real numbers from -2 to 2 inclusive
Range: real numbers from -3 to 9 inclusive

Chapter 15 Practice Test

1.
$$3x^2 = 42$$
$$3x^2 - 42 = 0$$
$a = 3; b = 0; c = -42$
pure

3.
$$5x^2 = 7x$$
$$5x^2 - 7x = 0$$
$a = 5; b = -7; c = 0$
incomplete

5.
$$x^2 = 81$$
$$x = \pm\sqrt{81}$$
$$x = \pm 9$$
$$x = 9 \text{ or } x = -9$$

7. $3x^2 - 6x = 0$

$3x(x - 2) = 0$

$3x = 0$ or $x - 2 = 0$

$\dfrac{3x}{3} = \dfrac{0}{3} \qquad x = 2$

$x = 0$

$x = 0$ or $x = 2$

9. $x^2 - 5x + 6 = 0$

$(x - 2)(x - 3) = 0$

$x - 2 = 0$ or $x - 3 = 0$

$\qquad x = 2 \quad$ or $\quad x = 3$

11.
$2x^2 + 12 = 11x$

$2x^2 - 11x + 12 = 0$

$2x^2 - 3x - 8x + 12 = 0$

$(2x^2 - 3x) + (-8x + 12) = 0$

$x(2x - 3) - 4(2x - 3) = 0$

$(2x - 3)(x - 4) = 0$

$2x - 3 = 0$ or $x - 4 = 0$

$\quad 2x = 3 \qquad x = 4$

$\qquad x = \dfrac{3}{2}$

$\qquad x = \dfrac{3}{2}$ or $x = 4$

$2 \cdot 12 = \quad 24$

$\overline{\qquad\quad 1 \cdot 24}$

$\qquad\quad 2 \cdot 12$

$\bigstar \; 3 \cdot 8 \quad (-3 - 8 = -11)$

$\qquad\quad 4 \cdot 6$

13. $2x^2 + 3x - 5 = 0$

$a = 2; \; b = 3; \; c = -5$

$x = \dfrac{-b \pm \sqrt{b^2 - 4ac}}{2a}$

$x = \dfrac{-3 \pm \sqrt{3^2 - 4(2)(-5)}}{2(2)}$

$x = \dfrac{-3 \pm \sqrt{9 + 40}}{4}$

$x = \dfrac{-3 \pm \sqrt{49}}{4}$

$x = \dfrac{-3 \pm 7}{4}$

$x = \dfrac{-3 + 7}{4} \qquad$ or $\qquad x = \dfrac{-3 - 7}{4}$

$x = \dfrac{4}{4} \qquad\qquad\qquad x = \dfrac{-10}{4}$

$x = 1 \qquad\quad$ or $\qquad x = \dfrac{-5}{2}$ or -2.5

15. $x^2 - 3x - 5 = 0$

$a = 1; \; b = -3; \; c = -5$

$x = \dfrac{-b \pm \sqrt{b^2 - 4ac}}{2a}$

$x = \dfrac{-(-3) \pm \sqrt{(-3)^2 - 4(1)(-5)}}{2(1)}$

$x = \dfrac{3 \pm \sqrt{9 + 20}}{2}$

$x = \dfrac{3 \pm \sqrt{29}}{2}$

$x = \dfrac{3 + \sqrt{29}}{2}$ or $x = \dfrac{3 - \sqrt{29}}{2} \qquad$ exact solutions

$x \approx 4.19 \quad$ or $\quad x \approx -1.19 \qquad$ approximate solutions

17. $x^2 - 8x + 12 = 0$

Examine the discriminant: $b^2 - 4ac$

$a = 1; b = -8; c = 12$

$b^2 - 4ac = (-8)^2 - 4(1)(12)$

$\qquad\qquad = 64 - 48$

$\qquad\qquad = 16$

Since 16 is a perfect square,
the two roots of the equation
are real, rational, and unequal.

19. $y = x^2 + 2x + 1$

x-coordinate of the vertex: $\dfrac{-b}{2a} = \dfrac{-2}{2(1)} = \dfrac{-2}{2} = -1$ $\quad$ $(-1, 0)$

$0 = x^2 + 2x + 1$

$0 = (x + 1)^2$

$x = -1$ (double root and x-intercept)

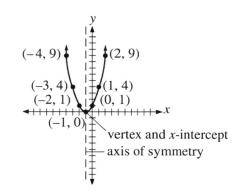

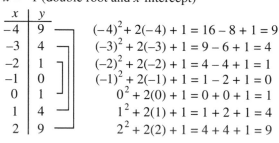

x	y
-4	9	$(-4)^2 + 2(-4) + 1 = 16 - 8 + 1 = 9$
-3	4	$(-3)^2 + 2(-3) + 1 = 9 - 6 + 1 = 4$
-2	1	$(-2)^2 + 2(-2) + 1 = 4 - 4 + 1 = 1$
-1	0	$(-1)^2 + 2(-1) + 1 = 1 - 2 + 1 = 0$
0	1	$0^2 + 2(0) + 1 = 0 + 0 + 1 = 1$
1	4	$1^2 + 2(1) + 1 = 1 + 2 + 1 = 4$
2	9	$2^2 + 2(2) + 1 = 4 + 4 + 1 = 9$

21.
$$R = \frac{KL}{d^2}$$
$$1.314 = \frac{(10.4)(3{,}642.5)}{d^2}$$
$$1.314(d^2) = (d^2)\frac{(10.4)(3{,}642.5)}{d^2}$$
$$1.314d^2 = 37{,}882$$
$$\frac{1.314d^2}{1.314} = \frac{37{,}882}{1.314}$$
$$d^2 = 28{,}829.53$$
$$d = \pm\sqrt{28{,}829.532816}$$
$$d = 169.79 \text{ mils}$$
Disregard negative root.

23.
$$R = \frac{W}{I^2}$$
$$52.29 = \frac{205}{I^2}$$
$$(I^2)52.29 = (I^2)\frac{205}{I^2}$$
$$52.29I^2 = 205$$
$$\frac{52.29I^2}{52.29} = \frac{205}{52.29}$$
$$I^2 = 3.920443679$$
$$I = \sqrt{3.920443679}$$
$$I = 1.98 \text{ amps}$$
Disregard negative root.

25.
$$E = 0.5mv^2$$
$$180 = 0.5(10)v^2$$
$$180 = 5v^2$$
$$\frac{180}{5} = \frac{5v^2}{5}$$
$$36 = v^2$$
$$\pm\sqrt{36} = v$$
$$v = 6$$
Disregard negative root.

27. $6x^3 + 21x^2 = 45x$

$6x^3 + 21x^2 - 45x = 0$ $\qquad$ $2 \cdot 15 = \dfrac{30}{3 \cdot 10}$

$3x(2x^2 + 7x - 15) = 0$ $\qquad$ ★ $(-3 + 10 = +7)$

Factor trinomial:

$\quad (2x^2 - 3x + 10x - 15)$

$\quad (2x^2 - 3x) + (10x - 15)$

$\quad x(2x - 3) + 5(2x - 3)$

$\qquad 3x(2x - 3)(x + 5) = 0$

$\qquad\quad 3x = 0$ or $2x - 3 = 0$ or $x + 5 = 0$

$\qquad\quad \dfrac{3x}{3} = \dfrac{0}{3} \qquad 2x = 3 \qquad\quad x = -5$

$\qquad\qquad x = 0 \qquad\quad \dfrac{2x}{2} = \dfrac{3}{2}$

$\qquad\qquad\qquad\qquad\quad x = \dfrac{3}{2}$

$\qquad\quad x = 0$ or $x = \dfrac{3}{2}$ or $x = -5$

29. $6x^3 - 18x^2 = 0$

$6x^2(x - 3) = 0$

$6x^2 = 0$ or $x - 3 = 0$

$x^2 = \dfrac{0}{6} \qquad\qquad x = 3$

$x^2 = 0$

$\sqrt{x^2} = \sqrt{0}$

$x = 0$ (double root)

$x = 0$ or $x = 3$

31. $x = y^3 - 2y^2 - y$
Domain: the set of all real numbers
Range: the set of all real numbers

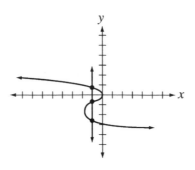

The relation is *not* a function.

Chapters 11–15 Cumulative Practice Test

1. $5x^7 \cdot 8x^3 = 5(8)x^{7+3} = 40x^{10}$

3. $\left(x^4\right)^8 = x^{4(8)} = x^{32}$

5. $5x^4 - 3x^2 + 7x^4 + 8x^2 =$
$5x^4 + 7x^4 - 3x^2 + 8x^2 = 12x^4 + 5x^2$

7. $\dfrac{10x^5 - 20x^6 + 15x^4}{5x^3} = \dfrac{10x^5}{5x^3} - \dfrac{20x^6}{5x^3} + \dfrac{15x^4}{5x^3} = 2x^2 - 4x^3 + 3x$

9. $24x^3 - 16x^2 - 8x = 8x\left(3x^2 - 2x - 1\right)$

11. $(4x - 5)(3x - 2) = 12x^2 - 8x - 15x + 10 = 12x^2 - 23x + 10$

13.
$$
\begin{array}{r}
x - 3 \\
x - 5 \overline{) x^2 - 8x + 15} \\
\underline{x^2 - 5x} \\
-3x + 15 \\
\underline{-3x + 15} \\
0
\end{array}
$$

15. $x^2 - 16x + 64 = (x - 8)^2$

17. $x^2 + x - 42 = (x + 7)(x - 6)$

19.
$$
\begin{aligned}
6x^2 + 17x + 12 &= \\
6x^2 + 8x + 9x + 12 &= \\
(6x^2 + 8x) + (+9x + 12) &= \\
2x(3x + 4) + 3(3x + 4) &= \\
(3x + 4)(2x + 3) &
\end{aligned}
$$

$6(12) = 72$
$1(72)$
$2(36)$
$3(24)$
$4(18)$
$6(12)$
$8(9)$ $\bigstar\, 8 + 9 = 17$

21. $3x^2 - 7 = 20$
$3x^2 = 20 + 7$
$3x^2 = 27$
$\dfrac{3x^2}{3} = \dfrac{27}{3}$
$x^2 = 9$
$x = \pm 3$
$x = 3$ or $x = -3$

23. $x^2 - 8x = 0$

$x(x - 8) = 0$

$x = 0$ or $x - 8 = 0$

$x = 0$ or $x = 8$

25. $2x^2 - 7x + 3 = 0$ $\quad 2(3) = 6$

$2x^2 - 1x - 6x + 3 = 0$ $\quad 1(6)$ $\star -1 - 6 = -7$

$(2x^2 - 1x) + (-6x + 3) = 0$ $\quad 2(3)$

$x(2x - 1) - 3(2x - 1) = 0$

$(2x - 1)(x - 3) = 0$

$2x - 1 = 0$ or $x - 3 = 0$

$x = \dfrac{1}{2}$ or $x = 3$

27. $4x^2 - 3x - 2 = 0$

$a = 4; \ b = -3; \ c = -2$

$x = \dfrac{-b \pm \sqrt{b^2 - 4ac}}{2a}$

$x = \dfrac{-(-3) \pm \sqrt{(-3)^2 - 4(4)(-2)}}{2(4)}$

$x = \dfrac{3 \pm \sqrt{9 + 32}}{8}$

$x = \dfrac{3 \pm \sqrt{41}}{8}$

$x \approx \dfrac{3 \pm 6.403124237}{8}$

$x \approx \dfrac{9.403124237}{8} \approx 1.18$ or $x \approx \dfrac{-3.403124237}{8} \approx -0.43$

29. $x^3 - 3x^2 - 10x = 0$

$x(x^2 - 3x - 10) = 0$

$x(x - 5)(x + 2) = 0$

$x = 0$ or $x - 5 = 0$ or $x + 2 = 0$

$\qquad\qquad x = 5 \qquad x = -2$

$x = 0$ or $x = 5$ or $x = -2$

chapter 16 Exponential and Logarithmic Equations

Chapter Review Exercises

1. $5^x = 5^8$
$x = 8$

3. $3^x = 3^{-2}$
$x = -2$

5. $4^{x-2} = 4^2$
$x - 2 = 2$
$x = 2 + 2$
$x = 4$

7. $6^{3x+2} = 6^{-3}$
$3x + 2 = -3$
$3x = -3 - 2$
$3x = -5$
$\dfrac{3x}{3} = \dfrac{-5}{3}$
$x = -\dfrac{5}{3} \text{ or } -1\dfrac{2}{3}$

9. $3^x = 81$
$3^x = 3^4$
$x = 4$

11. $2^x = \dfrac{1}{32}$
$2^x = 2^{-5}$
$x = -5$

13. $5^{3x} = 125$
$5^{3x} = 5^3$
$3x = 3$
$\dfrac{3x}{3} = \dfrac{3}{3}$
$x = 1$

15. $6^{2-x} = \dfrac{1}{36}$
$6^{2-x} = 6^{-2}$
$2 - x = -2$
$-x = -2 - 2$
$-x = -4$
$\dfrac{-x}{-1} = \dfrac{-4}{-1}$
$x = 4$

17. $e^{-4} = 0.01831563889$

some calculators: $\boxed{e^x}$ 4 $\boxed{^+\!/\!_-}$ $\boxed{=}$

other calculators: $\boxed{e^x}$ $\boxed{(-)}$ 4 $\boxed{\text{ENTER}}$

19. $e^{-10} = 0.00004539992976$

some calculators: $\boxed{e^x}$ 10 $\boxed{^+\!/\!_-}$ $\boxed{=}$;

other calculators: $\boxed{e^x}$ $\boxed{(-)}$ 10 $\boxed{\text{ENTER}}$

21. $I = P(1 + R)^N - P$

     $I = 1{,}600(1 + 0.13)^3 - 1{,}600$ $P = \$1{,}600$

     $I = 1{,}600(1.13)^3 - 1{,}600$ $R = 0.13$

     $I = 1{,}600(1.442897) - 1{,}600$ $N = 3$

     $I = 2{,}308.6352 - 1{,}600$

     $I = \$708.64$

     Calculator: 1600 $\boxed{(}$ 1 $\boxed{\cdot}$ 13 $\boxed{)}$ $\boxed{\wedge}$ 3 $\boxed{-}$ 1600 $\boxed{\text{ENTER}}$ $\Rightarrow$ 708.6352

23. $I = P(1 + R)^N - P$

     $I = 2{,}000(1 + 0.06)^4 - 2{,}000$ $P = \$2{,}000$

     $I = 2{,}000(1.06)^4 - 2{,}000$ $R = \dfrac{0.12}{2} = 0.06$

     $I = 2{,}000(1.26247696) - 2{,}000$

     $I = 2{,}524.95392 - 2{,}000$ $N = 2(2) = 4$

     $I = \$524.95$

     Calculator: 2000 $\boxed{(}$ 1 $\boxed{\cdot}$ 06 $\boxed{)}$ $\boxed{\wedge}$ 4 $\boxed{-}$ 2000 $\boxed{\text{ENTER}}$ $\Rightarrow$ 524.95392

25. $A = P(1 + R)^N$ $P = \$3{,}000$

     $A = 3{,}000(1 + 0.06)^{10}$ $R = \dfrac{0.12}{2} = 0.06$

     $A = 3{,}000(1.06)^{10}$ $N = 5(2) = 10$

     $A = 3{,}000(1.790847697)$

     $A = \$5{,}372.54$ Rounding could differ slightly from text answer.

27. $I = P(1 + R)^N - P$ $P = \$10{,}000$

     $I = 10{,}000(1 + 0.02)^{10} - 10{,}000$ $R = \dfrac{0.04}{2} = 0.02$

     $I = 10{,}000(1.02)^{10} - 10{,}000$ $N = 5(2) = 10$

     $I = 10{,}000(1.21899442) - 10{,}000$

     $I = 12{,}189.9442 - 10{,}000$

     $I = \$2{,}189.94$ Rounding could differ slightly from text answer.

29. $FV = PV(1 + R)^N$ $PV = \$8{,}000$

     $FV = 8{,}000(1 + 0.02)^{28}$ $R = \dfrac{0.08}{4} = 0.02$

     $FV = 8{,}000(1.02)^{28}$ $N = 7(4) = 28$

     $FV = 8{,}000(1.741024206)$

     $FV = \$13{,}928.19$ Rounding could differ slightly from text answer.

31. $A = Pe^{rt}$ $\qquad$ $P = \$5,000$

$A = 5,000e^{(0.12)2}$ $\qquad$ $t = 2$ years

$A = 5,000e^{0.24}$ $\qquad$ $r = 0.12$

$A = 5,000(1.27124915)$

$A = \$6,356.25$

$I = A - P$

$I = \$6,356.25 - \$5,000 = \$1,356.25$

33. $E = \left(1 + \dfrac{r}{n}\right)^{n} - 1$

$E = \left(1 + \dfrac{0.08}{4}\right)^{4} - 1$

$E = (1 + 0.02)^{4} - 1$

$E = (1.02)^{4} - 1$

$E = 0.08243216$

$E = 8.24\%$

35. $I = P(1 + R)^{N} - P$ $\qquad$ $P = \$1,500$

$I = 1,500(1 + 0.0025)^{48} - 1,500$ $\qquad$ $R = \dfrac{0.03}{12} = 0.0025$

$I = 1,500(1.0025)^{48} - 1,500$ $\qquad$ $N = 4(12) = 48$

$I = 1,500(1.127328021) - 1,500$

$I = 1,690.9920316 - 1,500$

$I = \$190.99$

37. $PV = \dfrac{FV}{(1 + R)^{N}}$ $\qquad$ $FV = \$4,000$

$PV = \dfrac{4,000}{(1 + 0.02)^{16}}$ $\qquad$ $R = \dfrac{0.08}{4} = 0.02$

$PV = \dfrac{4,000}{(1.02)^{16}}$ $\qquad$ $N = 4(4) = 16$

$PV = \dfrac{4,000}{1.372785705}$

$PV = \$2,913.78$

39. $PV = \dfrac{FV}{(1 + R)^{N}}$ $\qquad$ $FV = \$11,000$

$PV = \dfrac{11,000}{(1 + 0.01)^{18}}$ $\qquad$ $R = \dfrac{0.12}{12} = 0.01$

$PV = \dfrac{11,000}{(1.01)^{18}}$ $\qquad$ $N = 1.5(12) = 18$

$PV = \dfrac{11,000}{1.196147476}$

$PV = \$9,196.19$ The collector should sell the painting to an individual for
$\$11,000$ to be paid in 18 months. Its present value is $\$9,196.19$, which is more than $\$8,000$.

41. $PV = \dfrac{FV}{(1 + R)^{N}}$ $\qquad$ $FV = \$7,000$

$PV = \dfrac{7,000}{(1 + 0.1)^{4}}$ $\qquad$ $R = 0.1$

$PV = \dfrac{7,000}{(1.1)^{4}}$ $\qquad$ $N = 4$

$PV = \dfrac{7,000}{1.4641}$

$PV = \$4,781.09$

43. $FV = P\left[\dfrac{(1 + R)^{N} - 1}{R}\right]$ $\qquad$ $P = \$2,000$

$FV = 2,000\left[\dfrac{(1 + 0.06)^{10} - 1}{0.06}\right]$ $\qquad$ $R = \dfrac{0.12}{2} = 0.06$

$FV = 2,000\left[\dfrac{(1.06)^{10} - 1}{0.06}\right]$ $\qquad$ $N = 5(2) = 10$

$FV = 2,000\left[\dfrac{0.7908476965}{0.06}\right]$

$FV = \$2,000(13.18079494)$

$FV = \$26,361.59$

45. $P = FV\left[\dfrac{R}{(1+R)^N - 1}\right]$　　　　$FV = \$155{,}000$

$P = 155{,}000\left[\dfrac{0.04}{(1+0.04)^{16} - 1}\right]$　　$R = \dfrac{0.08}{2} = 0.04$

$P = 155{,}000\left[\dfrac{0.04}{(1.04)^{16} - 1}\right]$　　　$N = 8(2) = 16$

$P = 155{,}000\left[\dfrac{0.04}{1.872981246}\right]$

$P = 155{,}000[0.0458199992]$

$P = \$7{,}102.10$

Calculator: 155000 $($ $\cdot$ 04 $\div$ $($ $($ 1 $+$ $\cdot$ 0 $)$ $\wedge$ 16 $-$ 1 $)$ $)$ ENTER

47. $P = FV\left[\dfrac{R}{(1+R)^N - 1}\right]$　　$FV = \$45{,}000$

$P = 45{,}000\left[\dfrac{0.01}{(1+0.01)^{18} - 1}\right]$　$R = \dfrac{0.12}{12} = 0.01$

$P = 45{,}000\left[\dfrac{0.01}{(1.01)^{18} - 1}\right]$　　$N = 1.5(12) = 18$

$P = 45{,}000\left[\dfrac{0.01}{0.196147476 - 1}\right]$

$P = 45{,}000\left[\dfrac{0.01}{0.196147476}\right]$

$P = 45{,}000[0.0509820479]$

$P = \$2{,}294.19$

49. $P = FV\left[\dfrac{R}{(1+R)^N - 1}\right]$　　$FV = \$75{,}000$

$P = 75{,}000\left[\dfrac{0.02}{(1+0.02)^{12} - 1}\right]$　$R = \dfrac{0.08}{4} = 0.02$

$P = 75{,}000\left[\dfrac{0.02}{(1.02)^{12} - 1}\right]$　　$N = 3(4) = 12$

$P = 75{,}000\left[\dfrac{0.02}{1.268241795 - 1}\right]$

$P = 75{,}000\left[\dfrac{0.02}{0.268241795}\right]$

$P = 75{,}000[0.0745595966]$

$P = \$5{,}591.97$

51. $P = FV\left[\dfrac{R}{(1+R)^N - 1}\right]$　　$FV = \$25{,}000$

$P = 25{,}000\left[\dfrac{0.08}{(1+0.08)^{6} - 1}\right]$　$R = 0.08$

$P = 25{,}000\left[\dfrac{0.08}{(1.08)^{6} - 1}\right]$　　$N = 6$

$P = 25{,}000\left[\dfrac{0.08}{1.586874323 - 1}\right]$

$P = 25{,}000\left[\dfrac{0.08}{0.5868743229}\right]$

$P = 25{,}000[0.1363153862]$

$P = \$3{,}407.88$

53. $M = P\left(\dfrac{R}{1 - (1+R)^{-N}}\right)$　　$P = \$238{,}000$

$M = 238{,}000\left(\dfrac{0.00625}{1 - (1+0.00625)^{-240}}\right)$　$R = \dfrac{0.075}{12} = 0.00625$

$M = 238{,}000\left(\dfrac{0.00625}{1 - (1.00625)^{-240}}\right)$　$N = 20(12) = 240$

$M = 238{,}000\left(\dfrac{0.00625}{1 - (0.22417418)}\right)$

$M = 238{,}000\left(\dfrac{0.00625}{0.77582582}\right)$

$M = 238{,}000(0.0080559319)$

$M = \$1{,}917.31$

55. $I = 1.50e^{-200t}$ $\quad\quad\quad t = 0.07$ s $\quad\quad$ **57.** $I = 1.50e^{-200t}$ $\quad\quad\quad t = 0.4$ s

$I = 1.50e^{-200(0.07)}$ $\quad\quad\quad\quad\quad\quad\quad\quad\quad I = 1.50e^{-200(0.4)}$

$I = 1.50e^{-14}$ $\quad\quad\quad\quad\quad\quad\quad\quad\quad\quad\quad I = 1.50e^{-80}$

$I = 0.000001247293079$ $\quad$ Calculator display: $\quad\quad I = 2.70727708 \times 10^{-35}$ $\quad$ Calculator display:

$I = 1.25 \times 10^{-6}$ $\quad\quad\quad\quad$ 1.247293079**E-6** $\quad\quad\quad\quad I = 2.71 \times 10^{-35}$ $\quad\quad\quad\quad$ 2.70727708**E-35**

59. $y = 5^x$

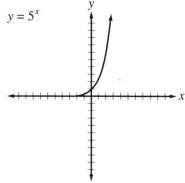

X	Y1	
−3	0.008	
−2	0.04	
−1	0.2	
0	1	
1	5	
2	25	
3	125	
X = −3		

61. $y = 5^{(x+4)}$

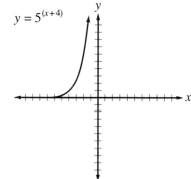

X	Y1	
−7	0.008	
−6	0.04	
−5	0.2	
−4	1	
−3	5	
−2	25	
−1	125	
X = −7		

63. $y = 5^{-x}$

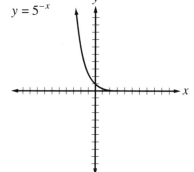

X	Y1	
−3	125	
−2	25	
−1	5	
0	1	
1	0.2	
2	0.04	
3	0.008	
X = −3		

65. $D = D_0(1 + r)^t$

$D = 20(1 - 0.04)^5$ $\quad$ Since the drug is breaking down, the rate 4% is negative.

$D = 20(0.96)^5$

$D = 16.3$ mg

67. $2^3 = 8$

$\log_2 8 = 3$

If $x = b^y$ then $\log_b x = y$.

69. $3^4 = 81$

$\log_3 81 = 4$

If $x = b^y$ then $\log_b x = y$.

71. $27^{1/3} = 3$

$\log_{27} 3 = \dfrac{1}{3}$

73. $4^{-3} = \dfrac{1}{64}$

$\log_4 \dfrac{1}{64} = -3$

75. $9^{-1/2} = \dfrac{1}{3}$

$\log_9 \dfrac{1}{3} = -\dfrac{1}{2}$

77. $12^{-2} = \dfrac{1}{144}$

$\log_{12} \dfrac{1}{144} = -2$

79. $\log_{11} 121 = 2$

$11^2 = 121$

$\log_b x = y$ converts to $x = b^y$.

81. $\log_{15} 1 = 0$

$15^0 = 1$

83. $\log_7 7 = 1$

$7^1 = 7$ or $7 = 7$

85. $\log_4 \dfrac{1}{16} = -2$

$4^{-2} = \dfrac{1}{16}$

87. $\log_9 \dfrac{1}{3} = -0.5$

$9^{-0.5} = \dfrac{1}{3}$

89. $\log_{10} 1,000 = 3$

$10^3 = 1,000$

$1,000 = 1,000$

91. $\log 5 = 0.6990$

calculator: $\boxed{\log}$ 5 $\boxed{\text{ENTER}}$

93. $\log 180 = 2.2553$

calculator: $\boxed{\log}$ 180 $\boxed{\text{ENTER}}$

95. $\log 0.4 = -0.3979$

calculator: $\boxed{\log}$ 0.4 $\boxed{\text{ENTER}}$

97. $\ln 270 = 5.5984$

calculator: $\boxed{\ln}$ 270 $\boxed{\text{ENTER}}$

99. $\ln 0.8 = -0.2231$

calculator: $\boxed{\ln}$ 0.8 $\boxed{\text{ENTER}}$

101. $\log_5 30 = \dfrac{\log 30}{\log 5} = \dfrac{1.4771}{0.6990} = 2.1133$

103. $\log_4 16 = x$

$4^x = 16$

$4^x = 4^2$

$x = 2$

105. $\log_7 x = 3$

$7^3 = x$

$343 = x$

107. $\log_6 \dfrac{1}{36} = x$

$6^x = \dfrac{1}{36}$

$6^x = 6^{-2}$

$x = -2$

109. $\log_2 x = 5$

$2^5 = x$

$32 = x$

111. Richter scale rating $= \log \dfrac{I}{I_0}$

(a) $R = \log \dfrac{100I_0}{I_0}$

$= \log 100$

$= 2$

(b) $R = \log \dfrac{10,000I_0}{I_0}$

$= \log 10,000$

$= 4$

(c) $R = \log \dfrac{150,000,000I_0}{I_0}$

$= \log 150,000,000$

$= 8.2$ (rounded)

113. $\log_2 9 = \log_2 3^2$

$= 2(\log_2 3)$

$= 2(1.585)$

$= 3.17$

115. $t = \dfrac{\ln A - \ln P}{r}$

$t = \dfrac{\ln 150,000 - \ln 100,000}{0.041}$

$t = \dfrac{11.91839057 - 11.51292546}{0.041}$

$t = \dfrac{0.405465105}{0.041}$

$t = 9.89$ years

$A = \$150,000$
$P = \$100,000$
$r = 4.1\%$

117. $y = \ln 4x$

X	Y1	
1	1.3863	
2	2.0794	
3	2.4849	
4	2.7726	
5	2.9957	
6	3.1781	
7	3.3322	
X = 1		

119. $y = \log 8x$

X	Y1	
1	0.9031	
2	1.2041	
3	1.3802	
4	1.5051	
5	1.6021	
6	1.6812	
7	1.7482	
X = 1		

121. $\text{pH} = 6.1 + \log\left(\dfrac{50}{c}\right)$

$7 = 6.1 + \log\left(\dfrac{50}{c}\right)$ $0.9 = \log\left(\dfrac{50}{c}\right)$ $10^{0.9} = \dfrac{50}{c}$ $10^{0.9}c = 50$ $c = \dfrac{50}{10^{0.9}}$ $c = 6.3$

Chapter 16 Practice Test

1. $1.2^{45} = 3657.26199$

calculator: $1.2 \boxed{y^x} \; 45 \boxed{=}$

or $1.2 \boxed{\wedge} 45 \boxed{\text{ENTER}}$

3. $15^{3/2} = 58.09475019$

calculator: $15 \boxed{y^x} \; 3 \boxed{\tfrac{b}{a/c}} \; 2 \boxed{=}$

or $15 \boxed{\wedge} \boxed{(}\; 3 \boxed{\div} 2 \boxed{)} \boxed{\text{ENTER}}$

5. $12^5 = 248{,}832$

calculator: $12 \boxed{y^x} \; 5 \boxed{=}$

or $12 \boxed{\wedge} 5 \boxed{\text{ENTER}}$

7. $2^{x-4} = 2^5$
$x - 4 = 5$
$x = 5 + 4$
$x = 9$

9. $2^{2x-1} = 8$
$2^{2x-1} = 2^3$
$2x - 1 = 3$
$2x = 3 + 1$
$2x = 4$
$\dfrac{2x}{2} = \dfrac{4}{2}$
$x = 2$

11. $4^{-1/2} = \dfrac{1}{2}$

$\log_4 \dfrac{1}{2} = -\dfrac{1}{2}$

13. $\log_3 \dfrac{1}{27} = -3$

$3^{-3} = \dfrac{1}{27}$

15. $\ln 32 = 3.4657$

calculator: $\boxed{\ln}\; 32 \boxed{\text{ENTER}}$

17. $\log_6 216 = x$
$6^x = 216$
$6^x = 6^3$
$x = 3$

19. $\log_8 21 = \dfrac{\log 21}{\log 8}$

$\log_8 21 = \dfrac{1.3222}{0.9031}$

$\log_8 21 = 1.4641$

21. $S = 125 + 83 \log(5t + 1)$
$= 125 + 83 \log(5(3) + 1)$
$= 125 + 83 \log(16)$
$= 125 + 83(1.20412)$
$= 125 + 99.94196$
$= \$224.942$ thousands
or $\$224{,}942$

23. $A = p\left(1 + \dfrac{r}{n}\right)^{nt}$

$p = \$5{,}000, \; r = 5.8\%, \; t = 2$ yrs, $n = 1$

$A = \; 5{,}000\left(1 + \dfrac{0.058}{1}\right)^{1(2)}$

$A = 5{,}000(1.058)^2$

$A = 5{,}000(1.119364)$

$A = \$5{,}596.82$

25. $FV = P\left[\dfrac{(1 + R)^N - 1}{R}\right]$ $\qquad P = \$9{,}000$
$\qquad\qquad\qquad\qquad\qquad\qquad\quad R = 0.15$
$FV = 9{,}000\left[\dfrac{(1 + 0.15)^2 - 1}{0.15}\right]$ $N = 2$

$FV = 9{,}000\left[\dfrac{(1.15)^2 - 1}{0.15}\right]$

$FV = 9{,}000\left[\dfrac{1.3225 - 1}{0.15}\right]$

$FV = 9{,}000\left[\dfrac{0.3225}{0.15}\right]$

$FV = \$9{,}000(2.15)$

$FV = \$19{,}350$

27. $P = FV\left[\dfrac{R}{(1+R)^N - 1}\right]$ $FV = \$125,000$
$R = 0.04$
$N = 16$

$P = 125,000\left[\dfrac{0.04}{(1+0.04)^{16} - 1}\right]$

$P = 125,000\left[\dfrac{0.04}{(1.04)^{16} - 1}\right]$

$P = 125,000\left[\dfrac{0.04}{1.872981246 - 1}\right]$

$P = 125,000\left[\dfrac{0.04}{0.872981246}\right]$

$P = 125,000[0.0458199992]$

$P = \$5,727.50$

29. $E = \left(1 + \dfrac{r}{n}\right)^n - 1$ $r = 0.12$

$E = \left(1 + \dfrac{0.12}{4}\right)^4 - 1$ $n = 4$

$E = (1 + 0.03)^4 - 1$

$E = (1.03)^4 - 1$

$E = 0.12550881$

$E = 12.55\%$

31. $A = P(1+R)^N$ $P = \$600$
$A = 600(1 + 0.01)^{12}$
$A = 600(1.01)^{12}$ $R = \dfrac{0.12}{12} = 0.01$
$A = 600(1.12682503)$ $N = 12$
$A = \$676.10$

$\$680$ in 1 year is slightly better than accepting $\$600$ now and investing it for 1 year.

33. Option 1
$A = P(1+R)^N$ $P = \$2,000$
$A = 2,000(1 + 0.02)^{16}$
$A = 2,000(1.02)^{16}$ $R = \dfrac{0.08}{4} = 0.02$
$A = 2,000(1.372785705)$
$A = \$2,745.57$ $N = 4(4) = 16$
Option 2
$A = P(1+R)^N$ $P = \$2,000$
$A = 2,000(1 + 0.0825)^4$ $R = 0.0825$
$A = 2,000(1.0825)^4$ $N = 4$
$A = 2,000(1.373129888)$
$A = \$2,746.26$
Option 2 yields slightly more interest.

35. $A = Pe^{rt}$ $P = \$1,000$
$A = 1,000e^{(0.04)20}$ $t = 20$ years
$A = 1,000e^{0.8}$ $r = 0.04$
$A - 1,000(2.225540928)$
$A = \$2,225.54$

37. $PV = \dfrac{FV}{(1+R)^N}$ $FV = \$15,000$
$R = 0.104$
$PV = \dfrac{15,000}{(1 + 0.104)^1}$ $N = 1$
$PV = \dfrac{15,000}{(1.104)^1}$
$PV = \$13,586.96$

chapter **17** | # Inequalities and Absolute Values

Chapter Review Exercises

1. The empty set is a set containing no elements.

$\{\ \ \}$ or $\emptyset$

3. $5 \in W$

5. $x > -7$

$(-7, \infty)$

7. $-4 \le x < 2$

$[-4, 2)$

9. $-2 < x$
or
$x > -2$

$(-2, \infty)$

11. $42 > 8m - 2m$
$42 > 6m$
$\dfrac{42}{6} > \dfrac{6m}{6}$
$7 > m$
$m < 7$

$(-\infty, 7)$

13. $0 < 2x - x$
$0 < x$
$x > 0$

$(0, \infty)$

15. $10 - 2x \ge 4$
$-2x \ge 4 - 10$
$-2x \ge -6$
$\dfrac{-2x}{-2} \le \dfrac{-6}{-2}$
$x \le 3$

$(-\infty, 3]$

17. $10x + 18 > 8x$
$18 > 8x - 10x$
$18 > -2x$
$\dfrac{18}{-2} < \dfrac{-2x}{-2}$
$-9 < x$
$x > -9$

$(-9, \infty)$

19. $12 + 5x > 6 - x$
$5x + x > 6 - 12$
$6x > -6$
$\dfrac{6x}{6} > \dfrac{-6}{6}$
$x > -1$

$(-1, \infty)$

21. $15 \ge 5(2 - y)$
$15 \ge 10 - 5y$
$15 - 10 \ge -5y$
$5 \ge -5y$
$\dfrac{5}{-5} \le \dfrac{-5y}{-5}$
$-1 \le y$
$y \ge -1$

$[-1, \infty)$

23. $6x - 2(x - 3) \le 30$

$6x - 2x + 6 \le 30$

$4x + 6 \le 30$

$4x \le 30 - 6$

$4x \le 24$

$\dfrac{4x}{4} \le \dfrac{24}{4}$

$x \le 6$

$(-\infty, 6]$

25. $D = P + \$5.60$

$2D + 12P \le \$59.80$

$2(P + 5.60) + 12P \le 59.80$

$2P + 11.20 + 12P \le 59.80$

$14P + 11.20 \le 59.80$

$14P \le 59.80 - 11.20$

$14P \le 48.60$

$\dfrac{14P}{14} \le \dfrac{48.60}{14}$

$P \le \$3.47$

$D \le 3.47 + 5.60$

$D \le \$9.07$

27. $A \cap B$ intersection: set that includes all elements that appear in *both* sets

{   }   no elements in common, empty set

29. $A \cap D$

$\{-1\}$

31. $(A \cup C)'$

$A \cup C = \{-2, -1, 0, 1, 2, 3, 4\}$

$(A \cup C)' = \{5\}$

33. $x + 4 > 2$

$x > 2 - 4$

$x > -2$

$(-2, \infty)$

35. $3x - 2 \le 4x + 1$

$3x - 4x \le 1 + 2$

$-x \le 3$

$\dfrac{-x}{-1} \ge \dfrac{3}{-1}$

$x \ge -3$

$[-3, \infty)$

37. $-3 < 2x - 4 < 5$

$-3 < 2x - 4$ and $2x - 4 < 5$

$-3 + 4 < 2x$ $\qquad$ $2x < 5 + 4$

$1 < 2x$ $\qquad$ $2x < 9$

$\dfrac{1}{2} < \dfrac{2x}{2}$ $\qquad$ $\dfrac{2x}{2} < \dfrac{9}{2}$

$\dfrac{1}{2} < x$ $\quad$ and $\quad$ $x < \dfrac{9}{2}$ or $4\dfrac{1}{2}$

$\dfrac{1}{2} < x < \dfrac{9}{2}$

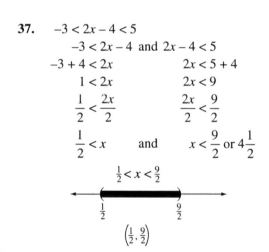

$\left(\dfrac{1}{2}, \dfrac{9}{2}\right)$

39. $4x + y < 2$

$y < -4x + 2$

$m = -4, \ b = 2$

dotted line

test: (0, 0)   test: (5, 0)

$4(0) + 0 < 2$ $\qquad$ $4(5) + 0 < 2$

$0 + 0 < 2$ $\qquad$ $20 + 0 < 2$

$0 < 2$ true $\qquad$ $20 < 2$ false

shade $\qquad\qquad$ do not shade

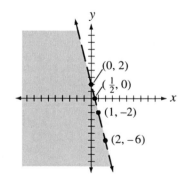

41. $3x + y \leq 2$
$$y \leq -3x + 2$$
$$m = -3, \ b = 2$$

solid line
test: $(0, 0)$ test: $(5, 0)$
$$3(0) + 0 \leq 2 \qquad\qquad 3(5) + 0 \leq 2$$
$$0 + 0 \leq 2 \qquad\qquad 15 + 0 \leq 2$$
$$0 \leq 2 \ \ \text{true} \qquad\qquad 15 \leq 2 \ \ \text{false}$$
$$\text{shade} \qquad\qquad\qquad \text{do not shade}$$

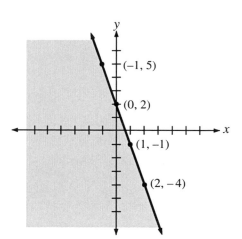

43. $x - 2y < 8$
$$-2y < -x + 8 \qquad \text{dotted line}$$
$$\frac{-2y}{-2} > \frac{-x}{-2} + \frac{8}{-2} \qquad \text{test: } (0, 0) \qquad\qquad\qquad \text{test: } (0, -5)$$
$$y > \frac{1}{2}x - 4 \qquad\qquad 0 - 2(0) < 8 \qquad\qquad\qquad 0 - 2(-5) < 8$$
$$0 - 0 < 8 \qquad\qquad\qquad 0 + 10 < 8$$
$$m = \frac{1}{2}, \ b = -4 \qquad\qquad 0 < 8 \ \ \text{true} \qquad\qquad\quad 10 < 8 \ \ \text{false}$$
$$\text{shade} \qquad\qquad\qquad\qquad \text{do not shade}$$

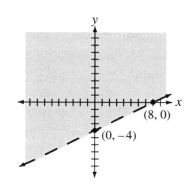

45. $y \geq 3x - 2$
$$m = 3, \ b = -2$$

solid line
test: $(0, 0)$ test: $(5, 0)$
$$0 \geq 3(0) - 2 \qquad\qquad 0 \geq 3(5) - 2$$
$$0 \geq 0 - 2 \qquad\qquad\quad 0 \geq 15 - 2$$
$$0 \geq -2 \ \ \text{true} \qquad\quad 0 \geq 13 \ \ \text{false}$$
$$\text{shade} \qquad\qquad\qquad \text{do not shade}$$

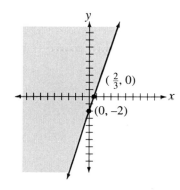

47. $y > \frac{2}{3}x - 2$
$$m = \frac{2}{3}, \ b = -2$$

dotted line
test: $(0, 0)$ test: $(6, 0)$
$$0 > \frac{2}{3}(0) - 2 \qquad\qquad 0 > \frac{2}{3}(6) - 2$$
$$0 > 0 - 2 \qquad\qquad\quad 0 > 4 - 2$$
$$0 > -2 \qquad \text{true} \qquad\quad 0 > 2 \qquad \text{false}$$
$$\text{shade} \qquad\qquad\qquad \text{do not shade}$$

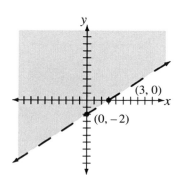

49. $2x + y < 6$
and
$x - y < 1$

$2x + y < 6$
$\quad y < -2x + 6$
$m = \dfrac{-2}{1}; \ b = 6$

dotted line
test: (0, 0)
$2(0) + 0 < 6$
$\quad 0 + 0 < 6$
$\qquad 0 < 6$ true
$\qquad$ shade
test: (5, 0)
$2(5) + 0 < 6$
$\quad 10 + 0 < 6$
$\qquad 10 < 6$ false
do not shade

$x - y < 1$
$\quad -y < -x + 1$
$\quad \dfrac{-y}{-1} > \dfrac{-x}{-1} + \dfrac{1}{-1}$
$\qquad y > x - 1$
$m = \dfrac{1}{1}; \ b = -1$

dotted line
test: (0, 0)
$0 - 0 < 1$
$\quad 0 < 1$ true
$\quad$ shade
test: (5, 0)
$5 - 0 < 1$
$\quad 5 < 1$ false
do not shade

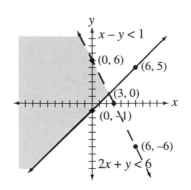

51. $2x + y > 3$
and
$x - y \le 1$

$2x + y > 3$
$\quad y > -2x + 3$
$m = \dfrac{-2}{1}; \ b = 3$

dotted line
test: (0, 0)
$2(0) + 0 > 3$
$\quad 0 + 0 > 3$
$\qquad 0 > 3$ false
do not shade

$x - y \le 1$
$\quad -y \le -x + 1$
$\quad \dfrac{-y}{-1} \ge \dfrac{-x}{-1} + \dfrac{1}{-1}$
$\qquad y \ge x - 1$
$m = \dfrac{1}{1}; \ b = -1$

solid line
test: (0, 0)
$0 - 0 \le 1$
$\quad 0 \le 1$ true
$\quad$ shade

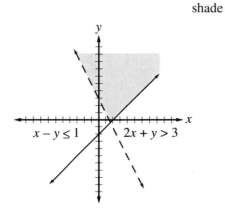

53. $x + 1 < 5 < 2x + 1$
$\quad x + 1 < 5 \qquad\qquad 5 < 2x + 1$
$\quad x < 5 - 1 \ \text{and} \ 5 - 1 < 2x$
$\quad x < 4 \qquad\qquad\quad 4 < 2x$
$\qquad\qquad\qquad\qquad \dfrac{4}{2} < \dfrac{2x}{2}$
$\qquad\qquad\qquad\qquad 2 < x$
$\qquad\quad 2 < x < 4$

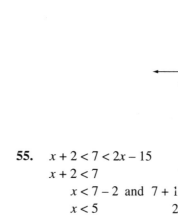

$\qquad\qquad (2, 4)$

55. $x + 2 < 7 < 2x - 15$
$\quad x + 2 < 7 \qquad\qquad 7 < 2x - 15$
$\quad x < 7 - 2 \ \text{and} \ 7 + 15 < 2x$
$\quad x < 5 \qquad\qquad\quad 22 < 2x$
$\qquad\qquad\qquad\qquad \dfrac{22}{2} < \dfrac{2x}{2}$
$\qquad\qquad\qquad\qquad 11 < x$
$\qquad$ NO SOLUTION

57. $2x + 3 < 15 < 3x + 9$

$2x + 3 < 15$ and $15 < 3x + 9$

$2x < 15 - 3 \quad 15 - 9 < 3x$

$2x < 12 \qquad\quad 6 < 3x$

$\dfrac{2x}{2} < \dfrac{12}{2} \qquad \dfrac{6}{3} < \dfrac{3x}{3}$

$x < 6 \qquad\quad 2 < x$

$2 < x < 6$

$(2, 6)$

59. $-5 \le -3x + 1 < 10$

$-5 \le -3x + 1$ and $-3x + 1 < 10$

$-5 - 1 \le -3x \qquad -3x < 10 - 1$

$-6 \le -3x \qquad\quad -3x < 9$

$\dfrac{-6}{-3} \ge \dfrac{-3x}{-3} \qquad \dfrac{-3x}{-3} > \dfrac{9}{-3}$

$2 \ge x \qquad\qquad x > -3$

$-3 < x \le 2$

$(-3, 2]$

61. $-3 \le 4x + 5 \le 2$

$-3 \le 4x + 5 \qquad 4x + 5 \le 2$

$-3 - 5 \le 4x \qquad\quad 4x \le 2 - 5$

$-8 \le 4x \qquad\qquad 4x \le -3$

$\dfrac{-8}{4} \le \dfrac{4x}{4} \qquad\quad \dfrac{4x}{4} \le \dfrac{-3}{4}$

$-2 \le x \qquad\qquad x \le \dfrac{-3}{4}$

$-2 \le x \le \dfrac{-3}{4}$

$\left[-2, -\dfrac{3}{4}\right]$

63. $x + 3 < 5 \quad \text{or} \quad x > 8$

$x < 2 \quad \text{or} \quad x > 8$

$(-\infty, 2) \cup (8, \infty)$

65. $x - 3 < -12 \qquad \text{or} \quad x + 1 > 9$

$x < -12 + 3 \qquad\quad x > 9 - 1$

$x < -9 \quad \text{or} \quad x > 8$

$(-\infty, -9) \cup (8, \infty)$

67. $x(1{,}365 + 199) > 15{,}000$

$x(1{,}564) > 15{,}000$

$x > \dfrac{15{,}000}{1{,}564}$

$x > 10 \text{ (rounded)}$

or

$x(1{,}365 + 199) < 30{,}000$

$x(1{,}564) < 30{,}000$

$x < \dfrac{30{,}000}{1{,}564}$

$x < 19 \text{ (rounded)}$

$(10, 19)$

69. $(x - 5)(x - 2) > 0$

critical values:

$x - 5 = 0$ ⠀⠀⠀⠀ $x - 2 = 0$

⠀⠀ $x = 5$ ⠀⠀⠀⠀⠀⠀ $x = 2$

$$\text{I} \mid \text{II} \mid \text{III}$$
$$\overleftrightarrow{\underset{2}{\circ} \quad \underset{5}{\circ}}$$

Region I	Region II	Region III
$x < 2$	$2 < x < 5$	$x > 5$
let $x = 1$	let $x = 3$	let $x = 6$
$(1 - 5)(1 - 2) > 0$	$(3 - 5)(3 - 2) > 0$	$(6 - 5)(6 - 2) > 0$
$(-4)(-1) > 0$	$(-2)(1) > 0$	$(1)(4) > 0$
$4 > 0$	$-2 > 0$	$4 > 0$
true	false	true

$$x < 2 \text{ or } x > 5$$

$$\overleftrightarrow{\underset{2}{)} \qquad \underset{5}{(}}$$

$$(-\infty, 2) \cup (5, \infty)$$

71. $(3x + 1)(2x - 3) < 0$

critical values: ⠀⠀⠀ $3x + 1 = 0$ ⠀⠀⠀ $2x - 3 = 0$

⠀⠀⠀⠀⠀⠀⠀⠀⠀⠀⠀⠀ $3x = -1$ ⠀⠀⠀⠀ $2x = 3$

$$\frac{3x}{3} = \frac{-1}{3} \qquad \frac{2x}{2} = \frac{3}{2}$$

$$x = \frac{-1}{3} \qquad x = \frac{3}{2} \text{ or } 1\frac{1}{2}$$

$$\text{I} \mid \text{II} \mid \text{III}$$
$$\overleftrightarrow{\underset{-\frac{1}{3}}{\circ} \quad \underset{1\frac{1}{2}}{\circ}}$$

Region I	Region II	Region III
$x < -\dfrac{1}{3}$	$-\dfrac{1}{3} < x < 1\dfrac{1}{2}$	$x > 1\dfrac{1}{2}$
let $x = -1$	let $x = 0$	let $x = 2$
$(3 \cdot -1 + 1)(2 \cdot -1 - 3) < 0$	$(3 \cdot 0 + 1)(2 \cdot 0 - 3) < 0$	$(3 \cdot 2 + 1)(2 \cdot 2 - 3) < 0$
$(-3 + 1)(-2 - 3) < 0$	$(0 + 1)(0 - 3) < 0$	$(6 + 1)(4 - 3) < 0$
$(-2)(-5) < 0$	$(1)(-3) < 0$	$(7)(1) < 0$
$10 < 0$	$-3 < 0$	$7 < 0$
false	true	false

$$-\frac{1}{3} < x < 1\frac{1}{2}$$

$$\overleftrightarrow{\underset{-\frac{1}{3}}{(} \qquad \underset{1\frac{1}{2}}{)}}$$

$$\left(-\frac{1}{3}, 1\frac{1}{2}\right)$$

73. $(x + 1)(x - 2) \leq 0$

critical values: $\quad x + 1 = 0 \qquad x - 2 = 0$

$$x = -1 \qquad x = 2$$

$$\text{I} \mid \text{II} \mid \text{III}$$
$$\xleftarrow{\quad} \underset{-1}{\ominus} + + \underset{2}{\ominus} \xrightarrow{\quad}$$

Region I	Region II	Region III
$x \leq -1$	$-1 \leq x \leq 2$	$x \geq 2$
let $x = -2$	let $x = 0$	let $x = 3$
$(-2 + 1)(-2 - 2) \leq 0$	$(0 + 1)(0 - 2) \leq 0$	$(3 + 1)(3 - 2) \leq 0$
$(-1)(-4) \leq 0$	$(1)(-2) \leq 0$	$(4)(1) \leq 0$
$4 \leq 0$	$-2 \leq 0$	$4 \leq 0$
false	true	false

$$-1 \leq x \leq 2$$
$$\xleftarrow{\quad} \underset{-1}{\blacksquare\!\!\!\blacksquare\!\!\!\blacksquare} \underset{2}{} \xrightarrow{\quad}$$
$$[-1, 2]$$

75. $2x^2 \leq 5x + 3$

$$2x^2 - 5x - 3 \leq 0 \qquad\qquad 2 \cdot 3 = \dfrac{6}{1 \cdot 6}$$
$$2x^2 + x - 6x - 3 \leq 0 \qquad\qquad\qquad +1 - 6 = -5$$
$$(2x^2 + x) + (-6x - 3) \leq 0$$
$$x(2x + 1) - 3(2x + 1) \leq 0$$
$$(2x + 1)(x - 3) < 0$$

critical values: $\quad 2x + 1 = 0 \qquad x - 3 = 0$

$$2x = -1 \qquad x = 3$$
$$x = \frac{-1}{2}$$

$$\text{I} \mid \text{II} \mid \text{III}$$
$$\xleftarrow{\quad} \underset{-\frac{1}{2}}{\ominus} + + \underset{3}{\ominus} \xrightarrow{\quad}$$

Region I	Region II	Region III
$x \leq -\dfrac{1}{2}$	$-\dfrac{1}{2} \leq x \leq 3$	$x \geq 3$
let $x = -1$	let $x = 0$	let $x = 4$
$2(-1)^2 \leq 5(-1) + 3$	$2(0)^2 \leq 5(0) + 3$	$2(4)^2 \leq 5(4) + 3$
$2(1) \leq -5 + 3$	$2(0) \leq 0 + 3$	$2(16) \leq 20 + 3$
$2 \leq -2$ false	$0 \leq 3$ true	$32 \leq 23$ false

$$-\frac{1}{2} \leq x \leq 3$$
$$\xleftarrow{\quad} \underset{-\frac{1}{2}}{\blacksquare\!\!\!\blacksquare\!\!\!\blacksquare} \underset{3}{} \xrightarrow{\quad}$$
$$\left[-\frac{1}{2}, 3 \right]$$

77. $2x^2 + 7x - 15 < 0$ $2 \cdot 15 = 30$ critical values:

$2x^2 - 3x + 10x - 15 < 0$ $\dfrac{}{1 \cdot 30}$ $2x - 3 = 0$ $x + 5 = 0$

$(2x^2 - 3x) + (10x - 15) < 0$ $2 \cdot 15$ $2x = 3$ $x = -5$

$x(2x - 3) + 5(2x - 3) < 0$ $3 \cdot 10 \;\star\; -3 + 10 = 7$ $\dfrac{2x}{2} = \dfrac{3}{2}$

$(2x - 3)(x + 5) < 0$

$$x = \frac{3}{2} \text{ or } 1\frac{1}{2}$$

I ┊   II   ┊ III

-5 $1\frac{1}{2}$

Region I	Region II	Region III
$x < -5$	$-5 < x < 1\dfrac{1}{2}$	$x > 1\dfrac{1}{2}$
let $x = -6$	let $x = 0$	let $x = 2$
$2(-6)^2 + 7(-6) - 15 < 0$	$2(0)^2 + 7(0) - 15 < 0$	$2(2)^2 + 7(2) - 15 < 0$
$72 - 42 - 15 < 0$	$0 + 0 - 15 < 0$	$8 + 14 - 15 < 0$
$15 < 0$ false	$-15 < 0$ true	$7 < 0$ false

$$-5 < x < 1\frac{1}{2}$$

-5 $1\frac{1}{2}$

$$\left(-5, 1\tfrac{1}{2}\right)$$

79. $\dfrac{x - 7}{x + 1} < 0;\; x = -1$ excluded value

critical values: $x - 7 = 0$ $x + 1 = 0$

                  $x = 7$ $x = -1$

I ┊   II $x = 7$ ┊ III

-1 7

$(-1, 7)$

Region I	Region II	Region III
$x < -1$	$-1 < x < 7$	$x > 7$
let $x = -2$	let $x = 0$	let $x = 8$
$\dfrac{-2 - 7}{-2 + 1} < 0$	$\dfrac{0 - 7}{0 + 1} < 0$	$\dfrac{8 - 7}{8 + 1} < 0$
$\dfrac{-9}{-1} < 0$	$\dfrac{-7}{1} < 0$	$\dfrac{1}{9} < 0$ false
$9 < 0$ false	$-7 < 0$ true	

$$-1 < x < 7$$

-1 7

$(-1, 7)$

81.
$$\frac{x}{x+8} > 0; \quad x = -8 \text{ excluded value}$$

critical values: $x = 0 \quad x + 8 = 0$

$$x = -8$$

$\oplus$	$\oplus$
-8	0

Region I	Region II	Region III
$x < -8$	$-8 < x < 0$	$x > 0$
let $x = -9$	let $x = -1$	let $x = 1$
$\dfrac{-9}{-9+8} > 0$	$\dfrac{-1}{-1+8} > 0$	$\dfrac{1}{1+8} > 0$
$\dfrac{-9}{-1} > 0$	$\dfrac{-1}{7} > 0$ false	$\dfrac{1}{9} > 0$ true
$9 > 0$ true		

$$x < -8 \quad \text{or} \quad x > 0$$

-8	0

$$(-\infty, -8) \cup (0, \infty)$$

83. $y \geq -2x^2$

vertex $\dfrac{-b}{2a} = \dfrac{-(0)}{2(-2)} = \dfrac{0}{-4} = 0$

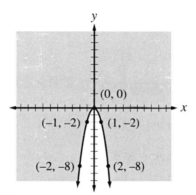

x	y	
2	-8	$-2(2)^2 = -2(4) = -8$
1	-2	$-2(1)^2 = -2(1) = -2$
0	0	$-2(0)^2 = -2(0) = 0$
-1	-2	$-2(-1)^2 = -2(1) = -2$
-2	-8	$-2(-2)^2 = -2(4) = -8$

solid parabola

test: (0, 5) outside   test: (0, −5) inside

$\quad 5 \geq -2(0)^2 \qquad\qquad -5 \geq -2(0)^2$

$\quad 5 \geq -2(0) \qquad\qquad\; -5 \geq -2(0)$

$\quad 5 \geq 0 \quad$ true $\qquad\; -5 \geq 0 \qquad$ false

$\quad$ shade

85. solid parabola

test: (3, 3) inside

$\quad 3 \geq (3)^2 - 6(3) + 9$

$\quad 3 \geq 9 - 18 + 9$

$\quad 3 \geq 0 \qquad$ true

test: (0, 0) outside

$\quad 0 \geq 0^2 - 6(0) + 9$

$\quad 0 \geq 9 \qquad$ false

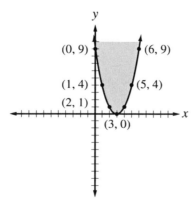

87. $|x| = 12$
$x = 12 \quad x = -12$

89. $|x + 3| = 7$
$x + 3 = 7 \qquad x + 3 = -7$
$x = 7 - 3 \qquad x = -7 - 3$
$x = 4 \qquad x = -10$

91. $|x - 8| = 12$
$x - 8 = 12 \qquad x - 8 = -12$
$x = 12 + 8 \qquad x = -12 + 8$
$x = 20 \qquad x = -4$

93. $|4x - 7| = 17$
$4x - 7 = 17 \qquad 4x - 7 = -17$
$4x = 17 + 7 \qquad 4x = -17 + 7$
$4x = 24 \qquad 4x = -10$
$\dfrac{4x}{4} = \dfrac{24}{4} \qquad \dfrac{4x}{4} = \dfrac{-10}{4}$
$x = 6$
$\qquad\qquad x = -\dfrac{5}{2} \text{ or } -2\dfrac{1}{2}$

95. $|7x + 8| = 15$
$7x + 8 = 15 \qquad 7x + 8 = -15$
$7x = 15 - 8 \qquad 7x = -15 - 8$
$7x = 7 \qquad 7x = -23$
$\dfrac{7x}{7} = \dfrac{7}{7} \qquad \dfrac{7x}{7} = \dfrac{-23}{7}$
$x = 1$
$\qquad\qquad x = \dfrac{-23}{7} \text{ or } -3\dfrac{2}{7}$

97. $|7x - 4| = 17$
$7x - 4 = 17 \qquad 7x - 4 = -17$
$7x = 17 + 4 \qquad 7x = -17 + 4$
$7x = 21 \qquad 7x = -13$
$\dfrac{7x}{7} = \dfrac{21}{7} \qquad \dfrac{7x}{7} = \dfrac{-13}{7}$
$x = 3$
$\qquad\qquad x = -\dfrac{13}{7} \text{ or } -1\dfrac{6}{7}$

99. $|3x - 9| = 2$
$3x - 9 = 2 \qquad 3x - 9 = -2$
$3x = 2 + 9 \qquad 3x = -2 + 9$
$3x = 11 \qquad 3x = 7$
$\dfrac{3x}{3} = \dfrac{11}{3} \qquad \dfrac{3x}{3} = \dfrac{7}{3}$
$\qquad\qquad\qquad x = \dfrac{7}{3} \text{ or } 2\dfrac{1}{3}$
$x = \dfrac{11}{3} \text{ or } 3\dfrac{2}{3}$

101. $|x| + 12 = 19$
$|x| = 19 - 12$
$|x| = 7$
$x = 7 \quad x = -7$

103. $|x| - 9 = 7$
$|x| = 7 + 9$
$|x| = 16$
$x = 16 \quad x = -16$

105. $-5 + |x - 3| = 2$
$|x - 3| = 2 - 5$
$|x - 3| = 7$
$x - 3 = 7 \qquad x - 3 = -7$
$x = 7 + 3 \qquad x = -7 + 3$
$x = 10 \qquad x = -4$

107. $|4x - 3| - 12 = -7$
$|4x - 3| = -7 + 12$
$|4x - 3| = 5$
$4x - 3 = 5 \qquad\qquad 4x - 3 = -5$
$4x = 5 + 3 \qquad\qquad 4x = -5 + 3$
$4x = 8 \qquad\qquad 4x = -2$
$\dfrac{4x}{4} = \dfrac{8}{4} \qquad\qquad \dfrac{4x}{4} = \dfrac{-2}{4}$
$x = 2$
$\qquad\qquad\qquad x = -\dfrac{1}{2}$

109.
$|x - 3| < 4$
$-4 < x - 3 < 4$
$-4 < x - 3 \quad x - 3 < 4$
$-4 + 3 < x \qquad x < 4 + 3$
$-1 < x \qquad\quad x < 7$
$-1 < x < 7$

$(-1, 7)$

111. $|x-4|-3<5$
$|x-4|<5+3$
$|x-4|<8$
$-8<x-4<8$
$-8<x-4 \quad x-4<8$
$-8+4<x \qquad x<8+4$
$-4<x \qquad x<12$
$-4<x<12$

$(-4, 12)$

113. $|x-3|<-4$
absolute value cannot equal negative value
NO SOLUTION

115. R = Riddle's income
S = Smith's income
D = Duke's income
$S<R<D$
$\$108,000 < R < \$250,000$

Chapter 17 Practice Test

1.

$x \geq -12$
-12
$[-12, \infty)$

3. $3x-1>8$
$3x>8+1$
$3x>9$
$\dfrac{3x}{3}>\dfrac{9}{3}$
$x>3$

$(3, \infty)$

5. $10<2+4x$
$10-2<4x$
$8<4x$
$\dfrac{8}{4}<\dfrac{4x}{4}$
$2<x$
$x>2$

$(2, \infty)$

7. $\dfrac{1}{3}x+5 \leq 3$
$(3)\dfrac{1}{3}x+(3)5 \leq (3)3$
$x+15 \leq 9$
$x \leq 9-15$
$x \leq 6$

-6
$(-\infty, -6]$

9. $5-3x<3-(2x-4)$
$5-3x<3-2x+4$
$5-3x<7-2x$
$-3x+2x<7-5$
$-x<2$
$\dfrac{-x}{-1}>\dfrac{2}{-1}$
$x>-2$

-2
$(-2, \infty)$

11. $-5<x+3<7$
$-5<x+3 \quad x+3<7$
$-5-3<x \qquad x<7-3$
$-8<x \qquad x<4$
$-8<x<4$

$-8 \qquad 4$
$(-8, 4)$

13. $3x-1 \leq 5 \leq x-5$
$3x-1 \leq 5 \qquad 5 \leq x-5$
$3x \leq 5+1 \qquad 5+5 \leq x$
$3x \leq 6 \qquad 10 \leq x$
$\dfrac{3x}{3} \leq \dfrac{6}{3} \qquad x \geq 10$
$x \leq 2$
NO SOLUTION

15. $(2x + 3)(x - 1) > 0$

critical values: $2x + 3 = 0$ $x - 1 = 0$

$2x = -3$ $x = 1$

$$\frac{2x}{2} = \frac{-3}{2}$$

$$x = -\frac{3}{2} \text{ or } -1\frac{1}{2}$$

$$\begin{array}{ccc} \text{I} & \text{II} & \text{III} \end{array}$$

$$\underset{-1\frac{1}{2} \qquad 1}{\longleftrightarrow}$$

Region I	Region II	Region III
$x < -1\dfrac{1}{2}$	$-1\dfrac{1}{2} < x < 1$	$x > 1$
let $x = -2$	let $x = 0$	let $x = 2$
$(2 \cdot -2 + 3)(-2 - 1) > 0$	$(2 \cdot 0 + 3)(0 - 1) > 0$	$(2 \cdot 2 + 3)(2 - 1) > 0$
$(-1)(-3) > 0$	$(3)(-1) > 0$	$(7)(1) > 0$
$+3 > 0$	$-3 > 0$	$7 > 0$
true	false	true

$$x < -1\tfrac{1}{2} \text{ or } x > 1$$

$$\underset{-1\frac{1}{2} \qquad 1}{\longleftrightarrow}$$

$$\left(-\infty, -1\tfrac{1}{2}\right) \cup (1, \infty)$$

17. $2x - 3 < 1$ or $x + 1 > 7$

$2x < 1 + 3$ $x + 1 > 7$

$2x < 4$ $x > 7 - 1$

$$\frac{2x}{2} < \frac{4}{2} \qquad x > 6$$

$$x < 2 \qquad \text{or} \qquad x > 6$$

$$\underset{2 \qquad\qquad 6}{\longleftrightarrow}$$

$$(-\infty, 2) \cup (6, \infty)$$

19. $A \cup B$ union: a set that includes all elements that appear in *either* set
$\{1, 2, 3, 4, 5, 6, 7, 8, 9\}$

21. $B' = \{9\}$ $A \cap B' = \{9\}$

23. $\dfrac{x-2}{x+5} < 0$

$x - 2 = 0 \quad x + 5 = 0$

$x = 2 \qquad x = -5 \quad$ critical values

I $\quad$ II $\quad$ III

$\quad$ −5 $\qquad$ 2

Region I $\qquad\qquad$ Region II $\qquad\qquad$ Region III

$x < -5 \qquad\qquad\quad -5 < x < 2 \qquad\qquad\quad x > 2$

let $x = -6 \qquad\qquad$ let $x = 0 \qquad\qquad\quad$ let $x = 3$

$\dfrac{-6-2}{-6+5} < 0 \qquad\quad \dfrac{0-2}{0+5} < 0 \qquad\qquad \dfrac{3-2}{3+5} < 0$

$\dfrac{-8}{-1} < 0 \qquad\qquad \dfrac{-2}{5} < 0 \text{ true} \qquad\quad \dfrac{1}{8} < 0 \text{ false}$

$8 < 0 \text{ false}$

$-5 < x < 2$

−5 $\qquad$ 2

$(-5, 2)$

25. $|x| = 15$

$x = 15 \quad x = -15$

27. $|x| + 8 = 10$

$|x| = 10 - 8$

$|x| = 2$

$x = 2 \qquad x = -2$

29. $|x + 8| > 10$

$x + 8 < -10 \quad \text{or} \quad x + 8 > 10$

$x < -10 - 8 \qquad\qquad x > 10 - 8$

$x < -18 \quad \text{or} \quad x > 2$

−18 $\qquad\qquad$ 2

$(-\infty, -18) \cup (2, \infty)$

31. $96.8 - 0.1 = 96.7$

$96.8 + 0.1 = 96.9$

Range of acceptable
measures: $96.7 < x \le 96.9$

33. $x + y < 4$ and $y > 3x + 2$

$x + y < 4 \qquad\qquad\qquad y > 3x + 2$

$\quad y < -x + 4 \qquad\qquad m = 3; \, b = 2$

$m = -1; \, b = 4$

dotted line $\qquad\qquad\qquad$ dotted line

test: $(0, 0) \qquad\qquad\qquad$ test: $(0, 0)$

$0 + 0 < 4 \qquad\qquad\qquad 0 > 3(0) + 2$

$\quad 0 < 4 \text{ true} \qquad\qquad 0 > 0 + 2$

$\qquad$ shade $\qquad\qquad\qquad 0 > 2 \quad$ false

test: $(5, 0) \qquad\qquad\qquad$ do not shade

$5 + 0 < 4$

$\quad 5 < 4 \text{ false} \qquad\qquad$ test: $(-5, 0)$

do not shade $\qquad\qquad\qquad 0 > 3(-5) + 2$

$\qquad\qquad\qquad\qquad\qquad 0 > -15 + 2$

$\qquad\qquad\qquad\qquad\qquad 0 > -13 \text{ true}$

$\qquad\qquad\qquad\qquad\qquad$ shade

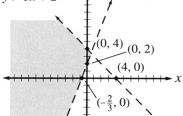

$y > 3x + 2$

$(0, 4)$ $\quad (0, 2)$

$(4, 0)$

$\left(-\dfrac{2}{3}, 0\right)$

$x + y < 4$

35. $2x - y \le 2$

$-y \le -2x + 2$

$\dfrac{-y}{-1} \ge \dfrac{-2x}{-1} + \dfrac{2}{-1}$

$y \ge 2x - 2$

$m = 2,\ b = -2$

solid line

test: $(0, 0)$

$2(0) - 0 \le 2$

$0 - 0 \le 2$

$0 \le 2$ true

shade

test: $(5, 0)$

$2(5) - 0 \le 2$

$10 - 0 \le 2$

$10 \le 2$ false

do not shade

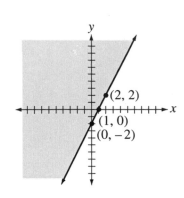

$(2, 2)$
$(1, 0)$
$(0, -2)$

37. $x + y < 1$

$y < -x + 1$

$m = -1,\ b = 1$

dotted line

test: $(0, 0)$

$0 + 0 < 1$

$0 < 1$ true

shade

test: $(5, 0)$

$5 + 0 < 1$

$5 < 1$ false

do not shade

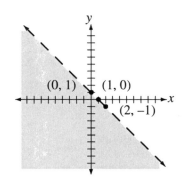

$(0, 1)$ $(1, 0)$
$(2, -1)$

39. $y \le x^2 - 6x + 8$

vertex $\dfrac{-b}{2a} = \dfrac{-(-6)}{2(1)} = \dfrac{6}{2} = 3$

solid parabola

$0 = x^2 - 6x + 8$

$0 = (x - 2)(x - 4)$

$x = 2$ or $x = 4$ solutions

$(2, 0)\ (4, 0)$ x-intercepts

x	y
6	8
5	3
4	0
3	−1
2	0
1	3
0	8

$6^2 - 6(6) + 8 = 36 - 36 + 8 = 8$

$5^2 - 6(5) + 8 = 25 - 30 + 8 = 3$

$4^2 - 6(4) + 8 = 16 - 24 + 8 = 0$

$3^2 - 6(3) + 8 = 9\ \ - 18 + 8 = -1$

$2^2 - 6(2) + 8 = 4\ \ - 12 + 8 = 0$

Use symmetry.

Use symmetry.

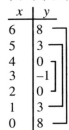

$y = 3^2 - 6(3) + 8$

$y = 9 - 18 + 8$

$(3, -1)$ $y = -1$

test: $(0, 0)$ outside

$0 \le 0^2 - 6(0) + 8$

$0 \le 0 - 0 + 8$

$0 \le 8$ true

shade

test: $(3, 0)$ inside

$0 \le 3^2 - 6(3) + 8$

$0 \le 9 - 18 + 8$

$0 \le -1$ false

do not shade

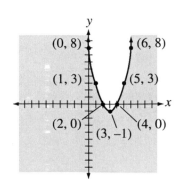

$(0, 8)$ $(6, 8)$
$(1, 3)$ $(5, 3)$
$(2, 0)$ $(4, 0)$
$(3, -1)$

41. $y < -\dfrac{1}{2}x^2$ $y = -\dfrac{1}{2}(0^2)$

$y = 0$

vertex $\dfrac{-b}{2a} = \dfrac{-(0)}{2\left(-\dfrac{1}{2}\right)} = \dfrac{0}{-1} = 0$ $(0, 0)$

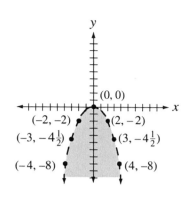

x	y
3	$-4\dfrac{1}{2}$
2	-2
1	$-\dfrac{1}{2}$
0	0
-1	$-\dfrac{1}{2}$
-2	-2
-3	$-4\dfrac{1}{2}$

$-\dfrac{1}{2}(3)^2 = -\dfrac{1}{2}(9) = -\dfrac{9}{2}$

$-\dfrac{1}{2}(2)^2 = -\dfrac{1}{2}(4) = -2$

$-\dfrac{1}{2}(1)^2 = -\dfrac{1}{2}(1) = -\dfrac{1}{2}$

$-\dfrac{1}{2}(0)^2 = -\dfrac{1}{2}(0) = 0$

$-\dfrac{1}{2}(-1)^2 = -\dfrac{1}{2}(1) = -\dfrac{1}{2}$

Use symmetry.

Use symmetry.

dotted parabola
test: $(0, 6)$ outside

$6 < -\dfrac{1}{2}(0)^2$

$6 < -\dfrac{1}{2}(0)$

$6 < 0$ false

do not shade

test: $(0, -5)$ inside

$-5 < -\dfrac{1}{2}(0)^2$

$-5 < -\dfrac{1}{2}(0)$

$-5 < 0$ true

shade

chapter 18   Geometry

Chapter Review Exercises

1. $\overleftrightarrow{AB} \parallel \overleftrightarrow{CD}$ **3.** $\overleftrightarrow{GH}$ intersects $\overleftrightarrow{EF}$ **5.** $\angle P$ **7.** $90°$; right

9. $180°$; straight **11.** $135° + 45° = 180°$; supplementary **13.** $21° + 79° = 100°$; neither

15. Two adjacent angles are supplementary when formed by intersecting lines.

$\angle a$ and $\angle b$ are supplementary angles.
$\angle b$ and $\angle c$ are supplementary angles.
$\angle c$ and $\angle d$ are supplementary angles.
$\angle d$ and $\angle a$ are supplementary angles.

17. Alternate exterior angles are on opposite sides of the transversal and are "outside" the parallel lines.

$\angle a$ and $\angle h$ are alternate exterior angles.
$\angle b$ and $\angle g$ are alternate exterior angles.

19. Angle d and $\angle f$ are supplementary angles.
$\angle f = 180° - 135°$
$\angle f = 45°$

21. $29'$ to decimal degree, ten-thousandth
$$29'\left(\frac{1°}{60'}\right) = 0.4833°$$

23. $7'\ 34''$
$$7'\ 34'' = 7' + 34''\left(\frac{1'}{60''}\right) = 7.5667'$$
$$= 7.5667'\left(\frac{1°}{60'}\right) = 0.1261°$$

25. $0.75°$ minutes and seconds
$$0.75° = 0.75°\left(\frac{60'}{1°}\right) = 45'$$

27. $0.2176°$ minutes and seconds
$$0.2176° = 0.2176(60') = 13.056'$$
$$= 13' + 0.056'\left(\frac{60''}{1'}\right)$$
$$= 13'\ 3''$$

29.

$P = 2(l + w)$
$= 2(65 + 40)$
$= 2(105)$
$= 210$ mm

31.

$P = b_1 + b_2 + s_1 + s_2$
 $= 7.5 + 14.3 + 6.8 + 7.2$
 $= 35.8$ in.

33.

$P = 2(b) + 2(s)$
 $= 2(18) + 2(15)$
 $= 36 + 30$
 $= 66$ ft

35.

$P = 2(l) + 2(w)$
 $= 2(84) + 2(60)$
 $= 168 + 120$
 $= 288$ in.

37.

$A = bh$
$A = 14(7)$
$A = 98$ cm^2

39.

$A = s^2$
$A = (5.9)^2$
$A = 34.81$ m^2

41.

$A = \dfrac{1}{2} bh$

$A = \dfrac{1}{2} (34)(18)$

$A = 306$ ft^2

or use Heron's formula

$s = \dfrac{1}{2} (a + b + c)$

$s = \dfrac{1}{2} (21.6 + 28.4 + 34)$

$s = \dfrac{1}{2} (84)$

$s = 42$

$A = \sqrt{s(s - a)(s - b)(s - c)}$

$A = \sqrt{42(42 - 21.6)(42 - 28.4)(42 - 34)}$

$A = \sqrt{42(20.4)(13.6)(8)}$

$A = \sqrt{93{,}219.84}$

$A = 305.3192428$

$A = 305.3$ ft^2

43. $A = lw$
$A = 300(120)$
$A = 36{,}000$ ft^2

45. $A = lw$

$A = 21 \left(18\dfrac{1}{2} \right)$

$A = 388.5$ ft^2

$A = \dfrac{388.5 \text{ ft}^2}{1} \left(\dfrac{1 \text{ yd}^2}{9 \text{ ft}^2} \right) \approx 43.16666667 \text{ yd}^2$

Round to the next whole yd^2 or 44 yd^2

47. $A_{\text{wall}} = lw$
    $= 25(11)$
    $= 275$ ft^2
$A_{\text{window}} = bh$
    $= 5(2)$
    $= 10$ ft^2
$A_{4 \text{ window}} = 4(10) = 40$ ft^2
$A_{\text{to stain}} = 275 - 40 = 235$ ft^2

49.

$A = \pi r^2$ $C = 2\pi r$
  $= \pi (4)^2$ $C = 2\pi (4)$
  $= 50.27$ m^2 $C = 8\pi$
              $C = 25.13$ m

51.

$A = lw$
$\quad = 4(3)$
$\quad = 12 \text{ cm}^2$

$P = \pi d + 2s$
$P = 3\pi + 2(4)$
$P = 17.42 \text{ cm}$

53. $C = \pi d$
$C = \pi(5)$
$C = 15.71 \text{ in.}$

$C = \dfrac{15.71 \text{ in.}}{1}\left[\dfrac{1 \text{ ft}}{12 \text{ in.}}\right] = 1.308996939 \text{ ft}$

Cutting speed $= C \times \text{rpm}$
$\qquad = 1.308996939(25)$
$\qquad = 32.72498347$
$\qquad = 33 \text{ ft/min}$

55. $60° = 60°\left[\dfrac{\pi \text{ rad}}{180°}\right] = \dfrac{\pi}{3} \text{ rad}$
$\qquad\qquad\quad$ or
$\qquad\qquad\quad = 1.05 \text{ rad}$

57. $99°45' = 99° + 45\left[\dfrac{1°}{60'}\right] = 99.75°$

$99.75° = 99.75°\left[\dfrac{\pi \text{ rad}}{180°}\right] = 1.74 \text{ rad}$

59. $\dfrac{5\pi}{6} \text{ rad}$

$\dfrac{5\pi}{6}\left[\dfrac{180°}{\pi}\right] = 150°$

61. $\theta = 45°9' = 45° + 9\left[\dfrac{1°}{60'}\right] = 45.15°$

$r = 2.58 \text{ cm}$

$A = \dfrac{\theta}{360}\pi r^2$

$A = \dfrac{45.15}{360}\pi(2.58)^2$

$A = 2.62 \text{ cm}^2 \text{ (rounded)}$

63. $\theta = 40°$

$r = 2\dfrac{1}{2} \text{ ft} = 2.5 \text{ ft}$

$A = \dfrac{\theta}{360}\pi r^2$

$A = \dfrac{40}{360}\pi(2.5)^2$

$A = 2.18 \text{ ft}^2 \text{ (rounded)}$

65. $\theta = 180°$

$r = 10 \text{ in.}$

$s = \dfrac{\theta}{360}2\pi r$

$s = \dfrac{180}{360}(2)\pi(10)$

$s = 31.42 \text{ in. (rounded)}$

67. $\theta = 30°$

$r = 24 \text{ cm}$

$h = 23.2 \text{ cm}$

$b = 12.4 \text{ cm}$

$A = \dfrac{\theta}{360}\pi r^2 - \dfrac{1}{2}bh$

$A = \dfrac{30}{360}\pi(24)^2 - \dfrac{1}{2}(12.4)(23.2)$

$A = 150.80 - 143.84$

$A = 6.96 \text{ cm}^2 \text{ (rounded)}$

69. $A = \dfrac{\theta}{360}\pi r^2 - \dfrac{1}{2}bh$

$A = \dfrac{106}{360}\pi(10)^2 - \dfrac{1}{2}(16)(6)$

$A = 92.50 - 48$

$A = 44.50 \text{ in}^2$

71.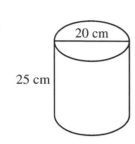

$$V = \pi(r)^2(h)$$
$$= \pi(10)^2(25)$$
$$= \pi(100)(25)$$
$$V = 7{,}854 \text{ cm}^3$$

$$r = \frac{d}{2} = \frac{20 \text{ cm}}{2}$$
$$= 10 \text{ cm}$$

73.
$$V = lwh$$
$$V_{\text{top soil}} = (85)(65)(0.5)$$
$$= 2{,}762.5 \text{ ft}^3$$

$$\frac{2{,}762.5 \text{ ft}^3}{1}\left(\frac{1 \text{ yd}^3}{27 \text{ ft}^3}\right) = 102 \text{ yd}^3$$

75.
$$TSA = ph + 2B$$
$$TSA = (27.8)(20) + 2(30)$$
$$TSA = 556 + 60$$
$$TSA = 616 \text{ cm}^2$$

$$p = 9 + 6.8 + 12$$
$$p = 27.8 \text{ cm}$$

$$B = \frac{1}{2}bh$$
$$B = \frac{1}{2}(12)(5)$$
$$B = 30 \text{ cm}^2$$
$$2B = 60 \text{ cm}^2$$

77.
$$TSA = ph + 2B$$
$$TSA = (94.24777961)(14) + 2(706.8583471)$$
$$= 1{,}319.468915 + 1{,}413.716694$$
$$= 2{,}733.1865609$$
$$= 2{,}733 \text{ cm}^2$$

$$p = C$$
$$C = 2\pi r$$
$$C = 2\pi(15)$$
$$C = 94.24777961 \text{ cm}$$
$$B = \pi r^2$$
$$= \pi(15)^2$$
$$= 706.8583471 \text{ cm}^2$$

79. cylinder, volume
diameter = 18 in.

$$18 \text{ in.} = 18 \text{ in.}\left(\frac{1 \text{ ft}}{12 \text{ in.}}\right) = 1.5 \text{ ft}$$

$$\text{height} = 5 \text{ mi} = \frac{5 \text{ mi}}{1}\left(\frac{5{,}280 \text{ ft}}{1 \text{ mi}}\right) = 26{,}400 \text{ ft}$$

$$LSA = ph \qquad p = \pi d$$
$$LSA = \pi dh$$
$$LSA = \pi(1.5)(26{,}400)$$
$$LSA = 124{,}407 \text{ ft}^2$$

81.
$$V = \frac{1}{3}Bh$$
$$V = \frac{1}{3}(12)(12)(36)$$
$$V = 1{,}728 \text{ m}^3$$

83.
$$TSA = 4\pi r^2$$
$$TSA = 4\pi(9)^2$$
$$TSA = 1{,}017.9 \text{ m}^2$$

85.
$$LSA = \pi rs$$
$$LSA = \pi(6)(9)$$
$$LSA = 169.6 \text{ cm}^2$$

87.
$$TSA = \pi rs + \pi r^2$$
$$TSA = \pi(9)(12) + \pi(9)^2$$
$$TSA = 339.3 + 254.5$$
$$TSA = 593.7610115 \text{ ft}^2$$
$$\frac{593.7610115 \text{ ft}^2}{1}\left(\frac{1 \text{ gal}}{350 \text{ ft}^2}\right) = 1.696460033 \text{ gallons}$$
Round to 2 gallons

89. largest angle is 80°, thus longest side is $\overline{ST}$
smallest angle is 42°, thus shortest side is $\overline{RS}$

Chapter 18 Practice Test

1. $0.3125° = 0.3125° \left(\dfrac{60'}{1°} \right)$

$= 18.75'$

$= 18' + 0.75' \left(\dfrac{60''}{1'} \right)$

$= 18'45''$

3. $35° = 35° \left(\dfrac{\pi \ rad}{180°} \right) = \dfrac{7\pi}{36}$ rad

or

$= 0.61$ rad

5. $\dfrac{5\pi}{8}$ rad $= \dfrac{5\pi}{8} \left(\dfrac{180°}{\pi} \right)$

$= 112.5°$

7. $P = 2(l + w)$
$= 2(32 \ \text{ft} \ 4 \ \text{in.} + 23 \ \text{ft} \ 2 \ \text{in.})$
$= 2(55 \ \text{ft} \ 6 \ \text{in.})$
$P = 111.0 \ \text{ft}$
$A = lw$

$= 32\dfrac{1}{3} \left(23\dfrac{1}{6} \right)$

$A = 749.1 \ \text{ft}^2$

9.

8 cm   12 cm

17 cm

$P = a + b + c$
$P = 8 + 12 + 17$
$P = 37$ cm

$s = \dfrac{1}{2}(a + b + c)$

$s = \dfrac{1}{2}(8 + 12 + 17)$

$s = 18.5$ cm

$A = \sqrt{s(s-a)(s-b)(s-c)}$

$A = \sqrt{18.5(18.5 - 8)(18.5 - 12)(18.5 - 17)}$

$A = \sqrt{18.5(10.5)(6.5)(1.5)}$

$A = \sqrt{1,893.9375}$

$A = 43.5 \ \text{cm}^2$

11.

23 m

$C = 2\pi r$
$C = 2\pi(23)$
$C = 46\pi$
$C = 144.5 \ \text{m}$
$A = \pi r^2$
$A = \pi (23)^2$
$A = \pi (529)$
$A = 1,661.9 \ \text{m}^2$

13. $\theta = 0.5$ $s = \theta r$
$r = 2$ in. $s = 0.5(2)$
$s = 1$ in.

15. $\theta = 1.7$ $s = \theta r$
$s = 2.9$ m $2.9 = 1.7r$

$\dfrac{2.9}{1.7} = r$

$r = 1.71$ m

17. $A = lw$
$11 \times 12 = 132 \ \text{ft}^2$
$10 \times 15 = 150 \ \text{ft}^2$
$\overline{\qquad 282 \ \text{ft}^2}$

19. $A_1 = 22.5 \ \text{ft} \ (30 \ \text{ft}) = 675 \ \text{ft}^2$
$A_2 = 23 \ \text{ft} \ (14.5 \ \text{ft}) = 333.5 \ \text{ft}^2$
$A_3 = 2 \ \text{ft} \ (4 \ \text{ft}) = 8 \ \text{ft}^2$
$A_1 + A_2 + A_3 = 675.5 \ \text{ft}^2 + 333.5 \ \text{ft}^2 + 8 \ \text{ft}^2$
$= 1,016.5 \ \text{ft}^2$

21. $A = \dfrac{\theta}{360} \pi r^2$

$A = \dfrac{42}{360} (\pi)(14 \ \text{cm})^2$

$A = 71.84 \ \text{cm}^2$

23. $LSA = ph$
$LSA = (5)(10)$
$LSA = 50 \ \text{in}^2$

pentagon, 5 sides, 1 in. each
perimeter $p = 5$ in.

25. $V = \dfrac{4\pi r^3}{3}$

$V = \dfrac{4\pi(6)^3}{3}$

$V = 904.8 \ \text{ft}^3$

where
$d = 12$ ft

$r = \dfrac{1}{2}d = \dfrac{1}{2}(12) = 6$ ft

$\dfrac{904.8 \ \text{ft}^3}{1} \left(\dfrac{7.48 \ \text{gal}}{1 \ \text{ft}^3} \right) = 6,768$ gal

Chapter Review Exercises

1. largest angle is $80°$, thus longest side is $\overline{ST}$
smallest angle is $42°$, thus shortest side is $\overline{RS}$

3. order of sides: 15, 13, 7.5
thus angles: $\angle C, \angle B, \angle A,$

5. $\angle B \cong \angle E$
$\angle C \cong \angle D$
$BC = DE$ or ED

7. $\angle J \cong \angle M$
$JL = MP$
$JK = MN$

9. $\dfrac{AB}{RT} = \dfrac{BC}{TS} = \dfrac{AC}{RS}$

11. $a = 9$ in. $c^2 = a^2 + b^2$
$b = 12$ in. $c^2 = 9^2 + 12^2$
$c = ?$ $c^2 = 81 + 144$
$\qquad\qquad c^2 = 225$
$\qquad\qquad c = \sqrt{225}$
$\qquad\qquad c = 15$ in.

13. $a = 7$ ft $c^2 = a^2 + b^2$
$b = ?$ $10^2 = 7^2 + b^2$
$c = 10$ ft $100 = 49 + b^2$
$\qquad\quad 100 - 49 = b^2$
$\qquad\qquad 51 = b^2$
$\qquad\qquad \sqrt{51} - b$
$\qquad\qquad\quad b = \sqrt{51}$ ft
$\qquad\qquad\qquad$ or
$\qquad\qquad\quad b = 7.141$ ft

15. $a = ?$ $c^2 = a^2 + b^2$
$b = 15$ yd $17^2 = a^2 + 15^2$
$c = 17$ yd $289 = a^2 + 225$
$\qquad\quad 289 - 225 = a^2$
$\qquad\qquad 64 = a^2$
$\qquad\qquad \sqrt{64} = a$
$\qquad\qquad\quad a = 8$ yd

17. $a = 11$ mi $c^2 = a^2 + b^2$
$b = 17$ mi $c^2 = 11^2 + 17^2$
$c = ?$ $c^2 = 121 + 289$
$\qquad\qquad c^2 = 410$
$\qquad\qquad c = \sqrt{410}$
$\qquad\qquad c = \sqrt{410}$ mi
$\qquad\qquad\qquad$ or
$\qquad\qquad c = 20.248$ mi

19. $a = ?$ $c^2 = a^2 + b^2$
$b = 40$ cm $50^2 = a^2 + 40^2$
$c = 50$ cm $2500 = a^2 + 1600$
$\qquad 2500 - 1600 = a^2$
$\qquad\qquad 900 = a^2$
$\qquad\qquad \sqrt{900} = a$
$\qquad\qquad\quad a = 30$ cm

21. $c^2 = a^2 + b^2$
$c^2 = 12^2 + 18^2$
$c^2 = 144 + 324$
$c^2 = 468$
$c = \sqrt{468}$
$c = 6\sqrt{13}$ in.
or
$c = 21.633$ in.

23.

$c^2 = a^2 + b^2$
$c^2 = 3^2 + 3^2$
$c^2 = 9 + 9$
$c^2 = 18$
$c = \sqrt{18}$
$c = 3\sqrt{2}$ in.
or
$c = 4.243$ in.

25. $RT = 15$ cm (hypotenuse)

$\text{leg} = \dfrac{\text{hyp}\sqrt{2}}{2}$

$\text{leg} = \dfrac{15\sqrt{2}}{2}$

$\text{leg} = 10.607$

$RS = ST = 10.607$ cm

27. $RT = 9\sqrt{2}$ hm (hypotenuse)

$\text{leg} = \dfrac{\text{hyp}\sqrt{2}}{2}$

$\text{leg} = \dfrac{9\sqrt{2} \cdot \sqrt{2}}{2} = \dfrac{9 \cdot 2}{2} = 9$

$RS = ST = 9$ hm

29. $AC = 12$ dm   (side opposite 60°)

$(\text{side opp. } 60°) = (\text{side opp. } 30°) \times \sqrt{3}$

$12 = (\text{side opp. } 30°) \times \sqrt{3}$

$\dfrac{12}{\sqrt{3}} = \text{side opp. } 30° = BC$

$6.928 \approx \text{side opp. } 30° \approx BC$

$AC^2 + BC^2 = AB^2$

$12^2 + (6.928)^2 \approx AB^2$

$144 + 47.997 \approx AB^2$

$191.997 \approx AB^2$

$\sqrt{191.997} \approx \sqrt{AB^2}$

$13.856 \approx AB$

$AB \approx 13.856$ cm

and $BC \approx 6.928$ cm

31. $BC = 10$ in.   (30° side)

$AB = \text{hyp} = (30°\text{side}) \times 2$

$AB = \text{hyp} = 10 \times 2$

$AB = \text{hyp} = 20$ in.

$AB = 20$ in.

$AC = 60°\text{side} = (30°\text{side}) \times \sqrt{3}$

$AC = 60°\text{side} = 10 \times \sqrt{3}$

$AC = 60°\text{side} = 17.321$

$AC = 17.321$ in.

33. $AC = 40$ ft 7 in. $= 40\dfrac{7}{12}$ ft or $\dfrac{487}{12}$ ft   (60°side)

$(60°\text{side}) = (30°\text{side}) \times \sqrt{3}$

$\dfrac{487}{12} = (30°\text{side}) \times \sqrt{3}$

$\dfrac{487}{12\sqrt{3}} = (30°\text{side}) = BC$

$23.43079842 = 30°\text{side} = BC$

$23.431 = 30°\text{side} = BC$

23 ft 5 in. $= 30°\text{side} = BC$

$AB = \text{hyp} = (30°\text{side}) \times 2$

$AB = \text{hyp} = 23.43079842 \times 2$

$AB = \text{hyp} = 46.86159685$

$AB = \text{hyp} = 46$ ft 10 in.

$AB = 46$ ft 10 in.

35.

$\text{hyp} = (30°\text{side}) \times 2$

$\text{hyp} = 14 \times 2$

$\text{hyp} = 28$ in.   (BC)

$AB + BC + CD = $ length of conduit

$(AB + CD) + BC$

71.751 in. + 28 in.

99.751 in.

100 in.

8 ft 4 in.

$60°\text{side} = (30°\text{side})\left(\sqrt{3}\right)$

$60°\text{side} = 14\sqrt{3}$

$60°\text{side} = 24.249$ in.   (BE)

$XY - BE = (AB + CD)$

8 ft – 24.249 in. $= (AB + CD)$

96 in. – 24.249 in. $= (AB + CD)$

71.751 in. $= (AB + CD)$

total length of conduit $ABCD$ is 100 in. or 8 ft 4 in.

37.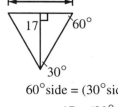

$60°\text{side} = (30°\text{side}) \times \sqrt{3}$

$17 = (30°\text{side}) \times \sqrt{3}$

$\dfrac{17}{\sqrt{3}} = 30°\text{side}$

$9.815 = 30°\text{side}$

$\text{width} = 2(9.815) = 19.630 \text{ mm}$

39. Area of $\Delta = \dfrac{1}{2}bh$

Area of $\Delta = \dfrac{1}{2}(23)(15.8)$

Area of $\Delta = 181.7 \text{ in}^2$

5Δ Area $= 5(181.7) = 908.5 \text{ in}^2$

41. regular pentagon

$\begin{aligned} \dfrac{\text{degrees of}}{\text{each angle}} &= \dfrac{180(n-2)}{n} \quad \text{where } n \text{ is number of sides} \\ &= \dfrac{180(5-2)}{5} \\ &= \dfrac{180(3)}{5} \\ &= 108° \end{aligned}$

43. $DO = 7$, $OB = ?$

$AO = 2(DO)$

$AO = 2(7)$

$AO = 14$

$OB = AO$

$OB = 14$

45. $AB = 10$, $AE = ?$

$AE = \dfrac{1}{2}AB = \dfrac{1}{2}(10)$

$AE = 5$

47. $\angle GJO = \dfrac{1}{2}\angle GJI$

$\angle GJO = \dfrac{1}{2}(90°)$

$\angle GJO = 45°$

49. $KO = 10$, $IJ = ?$

$KO = \dfrac{1}{2}\text{ side}$

$10 = \dfrac{1}{2}\text{ side}$

$2(10) = 2\left(\dfrac{1}{2}\text{ side}\right)$

$20 = \text{ side}$

$IJ = 20$

51. $\angle MOP = ?$

$\text{each angle} = \dfrac{180(n-2)}{n} = \dfrac{180(6-2)}{6}$

$\text{each angle} = \dfrac{180(4)}{6}$

$\text{each angle} = 120°$

$\angle LMN = 120°$

$\angle OMN = \dfrac{1}{2}\angle LMN = \dfrac{1}{2}(120°) = 60°$

$\angle OPM = 90°$

$\angle MOP = 30°$

53. $d = 20$ mm, thus $r = 10$ mm

$$r = \frac{2}{3}h$$

$$10 = \frac{2}{3}h$$

$$(3)10 = (\cancel{3})\frac{2}{\cancel{3}}h$$

$$30 = 2h$$

$$15 = h$$

height of Δ is 15 mm

$$r = \frac{1}{3}h$$

$$r = \frac{1}{3}(15)$$

$$r = 5 \text{ mm}$$

55.

$$\text{leg} = \frac{\text{hyp}\sqrt{2}}{2}$$

$$\text{leg} = \frac{5\sqrt{2}}{2}$$

$$\text{leg} = 3.54 \text{ cm}$$

57. $(3, 6)$ and $(-1, 4)$

$$d = \sqrt{(x_2 - x_1)^2 + (y_2 - y_1)^2}$$

$$d = \sqrt{(-1 - 3)^2 + (4 - 6)^2}$$

$$d = \sqrt{(-4)^2 + (-2)^2}$$

$$d = \sqrt{16 + 4}$$

$$d = \sqrt{20}$$

$$d = 4.472$$

59. $(3, -3)$ and $(0, 7)$

$$d = \sqrt{(x_2 - x_1)^2 + (y_2 - y_1)^2}$$

$$d = \sqrt{(0 - 3)^2 + (7 - (-3))^2}$$

$$d = \sqrt{(-3)^2 + 10^2}$$

$$d = \sqrt{9 + 100}$$

$$d = \sqrt{109}$$

$$d = 10.440$$

61. $(0, 0)$ and $(-3, 5)$

$$d = \sqrt{(x_2 - x_1)^2 + (y_2 - y_1)^2}$$

$$d = \sqrt{(-3 - 0)^2 + (5 - 0)^2}$$

$$d = \sqrt{(-3)^2 + 5^2}$$

$$d = \sqrt{9 + 25}$$

$$d = \sqrt{34}$$

$$d = 5.831$$

63. $(5, 2)$ and $(-3, -3)$

$$d = \sqrt{(x_2 - x_1)^2 + (y_2 - y_1)^2}$$

$$d = \sqrt{(-3 - 5)^2 + (-3 - 2)^2}$$

$$d = \sqrt{(-8)^2 + (-5)^2}$$

$$d = \sqrt{64 + 25}$$

$$d = \sqrt{89}$$

$$d = 9.434$$

65. (3, 6) and (−1, 4)

$$\text{midpoint} = \left(\frac{x_1 + x_2}{2}, \frac{y_1 + y_2}{2}\right)$$
$$= \left(\frac{3 + (-1)}{2}, \frac{6 + 4}{2}\right)$$
$$= \left(\frac{2}{2}, \frac{10}{2}\right)$$
$$\text{midpoint} = (1, 5)$$

67. (3, −3) and (0, 7)

$$\text{midpoint} = \left(\frac{x_1 + x_2}{2}, \frac{y_1 + y_2}{2}\right)$$
$$= \left(\frac{3 + 0}{2}, \frac{-3 + 7}{2}\right)$$
$$= \left(\frac{3}{2}, \frac{4}{2}\right)$$
$$\text{midpoint} = \left(1\frac{1}{2}, 2\right)$$

69. (0, 0) and (−3, 5)

$$\text{midpoint} = \left(\frac{x_1 + x_2}{2}, \frac{y_1 + y_2}{2}\right)$$
$$= \left(\frac{0 + (-3)}{2}, \frac{0 + 5}{2}\right)$$
$$= \left(\frac{-3}{2}, \frac{5}{2}\right)$$
$$\text{midpoint} = \left(-1\frac{1}{2}, 2\frac{1}{2}\right)$$

71. (5, 2) and (−3, −3)

$$\text{midpoint} = \left(\frac{x_1 + x_2}{2}, \frac{y_1 + y_2}{2}\right)$$
$$= \left(\frac{5 + (-3)}{2}, \frac{2 + (-3)}{2}\right)$$
$$= \left(\frac{2}{2}, -\frac{1}{2}\right)$$
$$\text{midpoint} = \left(1, -\frac{1}{2}\right)$$

73. (−5, −4) and (2, −2)

$$\text{midpoint} = \left(\frac{x_1 + x_2}{2}, \frac{y_1 + y_2}{2}\right)$$
$$= \left(\frac{-5 + 2}{2}, \frac{-4 + (-2)}{2}\right)$$
$$= \left(\frac{-3}{2}, \frac{-6}{2}\right)$$
$$\text{midpoint} = \left(-1\frac{1}{2}, -3\right)$$

Chapter 19 Practice Test

1.   largest $\angle A$
smallest $\angle B$

3.
$$\frac{9}{15} = \frac{12}{15 + x} \qquad \text{let } DB = x$$
$$9(15 + x) = 12(15)$$
$$9x + 135 = 180$$
$$9x = 180 - 135$$
$$9x = 45$$
$$\frac{9x}{9} = \frac{45}{9}$$
$$x = 5$$

$$DB = 5$$

5. $a = 20$ $c^2 = a^2 + b^2$
$b = 48$ $c^2 = 20^2 + 48^2$
$c = ?$ $c^2 = 400 + 2{,}304$
 $c^2 = 2{,}704$
 $c = 52$

7. $a = 8$ $c^2 = a^2 + b^2$
$b = 15$ $c^2 = 8^2 + 15^2$
$c = ?$ $c^2 = 64 + 225$
 $c^2 = 289$
 $c = 17$ in.

9. $AB = 62$ cm
$$AC = \frac{1}{2}(AB)$$
$$AC = \frac{1}{2}(62 \text{ cm})$$
$$AC = 31 \text{ cm}$$
$$BC = AC \; \sqrt{3}$$
$$BC = 31\sqrt{3} \text{ cm}$$
$$BC \approx 53.69 \text{ cm}$$

11. $AC = 5\dfrac{1}{2}$ cm
$$AC = \frac{11}{2} \text{ cm}$$
$$BC = AC$$
$$BC = \frac{11}{2} \text{ cm or } 5\frac{1}{2} \text{ cm}$$
$$AB = AC\sqrt{2}$$
$$AB = \frac{11\sqrt{2}}{2} \text{ cm}$$
$$AB \approx 7.78 \text{ cm}$$

13.

hyp $=$ leg $\sqrt{2}$
hyp $= 4\sqrt{2}$
hyp $= 5.657$ cm $= BC$

$ABEK = AB + BE + EK = 12$
if $BE = 4$
then $AB + EK = 8$
and $CD = EK$
thus $AB + CD = 8$
$ABCD = AB + BC + CD$
$ABCD = (AB + CD) + BC$
$ABCD = 8 + 5.657$
$ABCD = 13.657$ cm

15. $r = \dfrac{1}{3}h$ $C = 2\pi r$
$\phantom{r = \frac{1}{3}h}$ $C = 2\pi(4)$
$r = \dfrac{1}{3}(12)$ $C = 25.13$ in.
$r = 4$ in.

17. $d = \dfrac{1}{2}$ in.
$r = \dfrac{1}{4}$ in. $= 0.25$ in.
$C = 2\pi r$
$C = 2\pi(0.25)$
$C = 1.57$ in.

19. $(-2, 1)$ and $(3, 3)$

$$d = \sqrt{(x_2 - x_1)^2 + (y_2 - y_1)^2}$$
$$d = \sqrt{(3 - (-2))^2 + (3 - 1)^2}$$
$$d = \sqrt{5^2 + 2^2}$$
$$d = \sqrt{25 + 4}$$
$$d = \sqrt{29}$$
$$d = 5.385$$

$$\text{midpoint} = \left(\frac{x_1 + x_2}{2}, \frac{y_1 + y_2}{2}\right)$$
$$= \left(\frac{-2 + 3}{2}, \frac{1 + 3}{2}\right)$$
$$= \left(\frac{1}{2}, \frac{4}{2}\right)$$
$$\text{midpoint} = \left(\frac{1}{2}, 2\right)$$

Chapters 16 – 19 Cumulative Practice Test

1. $5^3 = 125$
  $\log_5 125 = 3$

3. $\log_6 36 = 2$
  $6^2 = 36$

5. $D = D_0(1 + r)^t$ $t = 8$ h, $D_0 = 300$ mg, $r = 0.12$
  $D = 300(1 - 0.12)^8$
  $D = 300(0.88)^8$
  $D = 300(0.3596345248)$
  $D = 107.8903574$
  $D = 107.9$ mg remain after 8 h.

7. $A = Pe^{rt}$ $P = \$12{,}000$
  $A = \$12{,}000e^{(0.045)(15)}$ $r = 0.045$
  $A = \$12{,}000e^{0.675}$ $t = 15$ years
  $A = \$23{,}568.40$

9. $5x - 3 < 7$
  $5x - 3 + 3 < 7 + 3$
  $5x < 10$
  $\dfrac{5x}{5} < \dfrac{10}{5}$
  $x < 2$ symbolic form

  graphical representation
  0 1 2 3 4

  $(-\infty, 2)$ interval notation

11. $-3 < x - 5 < 4$
  $-3 + 5 < x - 5 + 5 < 4 + 5$
  $2 < x < 9$ symbolic form

  graphical representation
  2   9

  $(2, 9)$ interval notation

13. $x + y \le 5$
  $x - y \ge 3$
  graph the equations:

Inequalities include boundary, so both lines are solid.

$x + y = 5$	$x - y = 3$
x-intercept	x-intercept
$x + 0 = 5$	$x + 0 = 3$
$x = 5$	$x = 3$
$(5, 0)$	$(3, 0)$
y-intercept	y-intercept
$0 + y = 5$	$0 - y = 3$
$y = 5$	$y = -3$
$(0, 5)$	$(0, -3)$

test: $(0, 0)$
$x + y \le 5$
$0 + 0 \le 5$ true
shade the side of the line
that contains $(0, 0)$

test: $(0, 0)$
$x - y \ge 3$
$0 - 0 \ge 3$ false
shade the side of the line
that does not contain $(0, 0)$

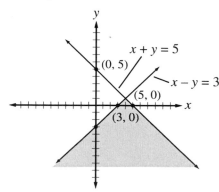

The solution is the shaded region shown on the graph.

15. $y \geq x^2 + x - 6$

graph $y = x^2 + x - 6$

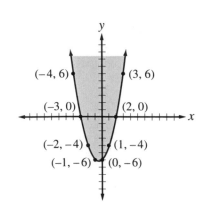

x	y
−4	6
−3	0
−2	−4
−1	−6
0	−6
1	−4
2	0
3	6

test point: (0, 0)

$y \geq x^2 + x - 6$

$0 \geq 0^2 + 0 - 6$

$0 \geq -6$ true

Shade portion that includes (0, 0)

The solution is the shaded portion.

The inequality includes the boundary ($\geq$); so the graph is a solid line.

17. acute (less than 90°)

19. straight (equal to 180°)

21. $90° - 28° = 62°$

23. $180° - 143° = 37°$

25. $30° + 12'\left(\dfrac{1°}{60'}\right) + 20''\left(\dfrac{1°}{3600''}\right) = 30° + 0.2° + 0.0056° = 30.2056°$

27. $0.352° = 0.352°\left(\dfrac{60'}{1°}\right) = 21.12'$

$21.12' = 21' + 0.12'\left(\dfrac{60''}{1'}\right) = 21' + 7.2''$ or $21'7''$ (rounded)

29. $P = a + b + c$

$P = 42 \text{ m} + 36 \text{ m} + 30 \text{ m}$

$P = 108 \text{ m}$

$s = \dfrac{1}{2}(a + b + c)$

$s = \dfrac{1}{2}(42 \text{ m} + 36 \text{ m} + 30 \text{ m})$

$s = \dfrac{1}{2}(180 \text{ m})$

$s = 54 \text{ m}$

$A = \sqrt{s(s - a)(s - b)(s - c)}$

$A = \sqrt{54(54 - 42)(54 - 36)(54 - 30)}$

$A = \sqrt{54(12)(18)(24)}$

$A = \sqrt{279{,}936}$

$A = 529 \text{ m}^2$ (rounded)

31. $C = \pi d$ $A = \pi r^2$ $r = \dfrac{d}{2}$

$C = \pi(150 \text{ cm})$ $A = \pi(75 \text{ cm})^2$

$C = 471 \text{ cm}$ $A = 17{,}671 \text{ cm}^2$ $r = \dfrac{150}{2} = 75 \text{ cm}$

33. $15°22' = 15° + 22'\left(\dfrac{1°}{60'}\right)$

$= 15° + 0.36666667°$

$= 15.36666667°$

$15.36666667°\left(\dfrac{\pi \text{ rad}}{180°}\right) = 0.2681989284 = 0.2682 \text{ rad}$ (rounded)

35. $1.8 \text{ rad}\left(\dfrac{180°}{\pi \text{ rad}}\right) = 103.1324031° = 103.1324°$

37. $V = \pi r^2 h$

$V = \pi(12 \text{ cm})^2(40 \text{ cm})$

$V = 18{,}095.57368$

$V = 18{,}096 \text{ cm}^3$

39.
$$c = \sqrt{a^2 + b^2}$$
$$c = \sqrt{(12 \text{ in.})^2 + (15 \text{ in.})^2}$$
$$c = \sqrt{144 + 225}$$
$$c = \sqrt{369}$$
$$c = 19.20937271$$
$$c = 19 \text{ in. (rounded)}$$

41.
$$AB = 32.5$$
$$AC = \frac{1}{2}(AB)$$
$$AC = \frac{1}{2}(32.5)$$
$$AC = 16.25 \text{ m}$$
$$BC = AC\sqrt{3}$$
$$BC = 16.25\sqrt{3}$$
$$BC = 28.1 \text{ m} \quad \text{rounded to the nearest tenth meter}$$

43. The diagonal of a square is the hypotenuse of each of the two triangles formed by the diagonal.

diagonal $= 38\sqrt{2}$

diagonal $= 54$ in. rounded to the nearest whole inch.

45.
$$d = \sqrt{(x_2 - x_1) + (y_2 - y_1)}$$
$$d = \sqrt{(9 - 3)^2 + (8 - 14)^2}$$
$$d = \sqrt{6^2 + (-6)^2}$$
$$d = \sqrt{36 + 36}$$
$$d = \sqrt{72}$$
$$d = 8.5$$

$x_1 = 3, \quad x_2 = 9,$
$y_1 = 14, \quad y_2 = 8$

chapter 20 Right-Triangle Trigonometry

Chapter Review Exercises

1.

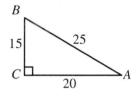

$$\sin A = \frac{\text{opp}}{\text{hyp}} = \frac{15}{25} = \frac{3}{5}$$

3.

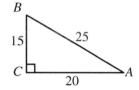

$$\tan A = \frac{\text{opp}}{\text{adj}} = \frac{15}{20} = \frac{3}{4}$$

5.

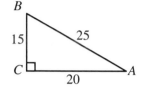

$$\sin B = \frac{\text{opp}}{\text{hyp}} = \frac{20}{25} = \frac{4}{5}$$

7.

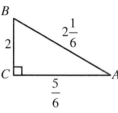

$$\sin A = \frac{\text{opp}}{\text{hyp}} = \frac{2}{2\frac{1}{6}}$$

$$= \frac{2}{1} \cdot \frac{6}{13} = \frac{12}{13}$$

$$10 \text{ in.} = \frac{10}{12} \text{ ft} = \frac{5}{6} \text{ ft} \qquad \text{or}$$

$$2 \text{ ft 2 in.} = 2 + \frac{2}{12} \text{ ft}$$

$$= 2\frac{1}{6} \text{ ft}$$

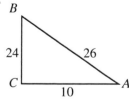

$$\sin A = \frac{24}{26} = \frac{12}{13}$$

$$2 \text{ ft 2 in.} = 2 \cdot 12 \text{ in.} + 2 \text{ in.}$$
$$= 24 \text{ in.} + 2 \text{ in.}$$
$$= 26 \text{ in.}$$

9.

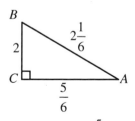

$$\tan B = \frac{\text{opp}}{\text{adj}} = \frac{\frac{5}{6}}{2}$$

$$= \frac{5}{6} \cdot \frac{1}{2} = \frac{5}{12}$$

$$10 \text{ in.} = \frac{10}{12} \text{ ft} = \frac{5}{6} \text{ ft}$$

$$2 \text{ ft 2 in.} = 2 + \frac{2}{12} \text{ ft}$$

$$= 2\frac{1}{6} \text{ ft}$$

or in inches

B

24   26

C ___10___ *A*

$$\tan B = \frac{\text{opp}}{\text{adj}} = \frac{10}{24} = \frac{5}{12}$$

11.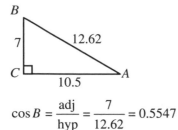

$$\cos B = \frac{\text{adj}}{\text{hyp}} = \frac{7}{12.62} = 0.5547$$

13.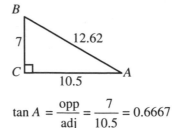

$$\tan A = \frac{\text{opp}}{\text{adj}} = \frac{7}{10.5} = 0.6667$$

15.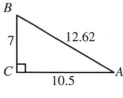

$$\tan B = \frac{\text{opp}}{\text{adj}} = \frac{10.5}{7} = 1.5$$

17. $\cos 32°50' = 0.8403$

19. $\sin 0.4712 = 0.4540$

21. $\tan 47° = 1.0724$

23. $\sin 0.8610 = 0.7585$

25. $\sin \theta = 0.5446$
$\theta = \sin^{-1}(0.5446)$
$\theta = 33.0°$

27. $\cos \theta = 0.6088$
$\theta = \cos^{-1}(0.6088)$
$\theta = 52.5°$

29. $\tan \theta = 2.723$
$\theta = \tan^{-1}(2.723)$
$\theta = 1.2188$ rad

31. $\tan \theta = 0.3440$
$\theta = \tan^{-1}(0.3440)$
$\theta = 0.3313$ rad

33.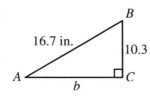

$$\sin A = \frac{\text{opp}}{\text{hyp}}$$
$$\sin A = \frac{10.3}{16.7}$$
$$\sin A = 0.6167664671$$
$$A = \sin^{-1}(0.6167664671)$$
$$A = 38.1°$$

35.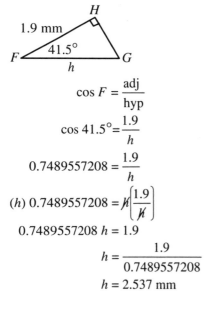

$$\cos F = \frac{\text{adj}}{\text{hyp}}$$
$$\cos 41.5° = \frac{1.9}{h}$$
$$0.7489557208 = \frac{1.9}{h}$$
$$(h)\,0.7489557208 = h\left(\frac{1.9}{h}\right)$$
$$0.7489557208\,h = 1.9$$
$$h = \frac{1.9}{0.7489557208}$$
$$h = 2.537 \text{ mm}$$

37.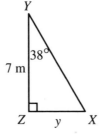

$$\tan Y = \frac{y}{x}$$
$$\tan 38° = \frac{y}{7}$$
$$y = 7 \tan 38°$$
$$y = 7(0.7812856265)$$
$$y = 5.468999386$$
$$y = 5.469 \text{ m}$$

39.

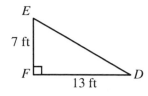

$$d^2 + e^2 = f^2$$
$$7^2 + 13^2 = f^2$$
$$49 + 169 = f^2$$
$$218 = f^2$$
$$\sqrt{218} = f$$
$$14.76 \text{ ft} = f$$

$$\tan D = \frac{\text{opp}}{\text{adj}}$$

$$\tan D = \frac{7}{13}$$

$$\tan D = 0.5384615385$$
$$D = \tan^{-1}(0.5384615385)$$
$$D = 28.3°$$

$$D + E = 90°$$
$$28.3° + E = 90°$$
$$E = 61.7°$$

41. Find A

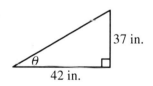

$$\sin A = \frac{8}{15}$$

$$\sin A = 0.5333333333$$
$$A = \sin^{-1}(0.5333333333)$$
$$A = 32.2°$$

43. Find a

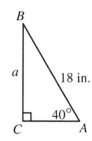

$$\sin 40° = \frac{a}{18}$$

$$0.6427876097 = \frac{a}{18}$$
$$(18)(0.6427876097) = (\cancel{18})\left(\frac{a}{\cancel{18}}\right)$$
$$11.57 \text{ in.} = a$$

45.

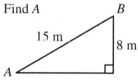

$$\tan \theta = \frac{37}{42}$$

$$\tan \theta = 0.880952381$$
$$\theta = \tan^{-1}(0.880952381)$$
$$\theta = 41.3785153°$$
$$\theta = 41.4°$$

two acute angles: $41.4°$, $48.6°$

$$90° - 41.4° = 48.6°$$

47.

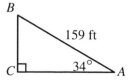

$$A + B = 90°$$
$$34° + B = 90°$$
$$B = 56°$$

$$\sin 34° = \frac{a}{159}$$

$$0.5591929035 = \frac{a}{159}$$
$$(159)(0.5591929035) = (\cancel{159})\left(\frac{a}{\cancel{159}}\right)$$
$$88.91167165 = a$$
$$88.91 \text{ ft} = a$$

$$a^2 + b^2 = c^2$$
$$(88.91167165)^2 + b^2 = 159^2$$
$$7905.285356 + b^2 = 25,281$$
$$b^2 = 17,375.71464$$
$$b = \sqrt{17,375.71464}$$
$$b = 131.816974$$
$$b = 131.8 \text{ ft}$$

49.

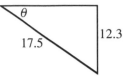

$$\sin \theta = \frac{\text{opp}}{\text{hyp}}$$

$$\sin \theta = \frac{12.3}{17.5}$$

$$\sin \theta = 0.7028571429$$

$$\theta = \sin^{-1}(0.7028571429)$$

$$\theta = 44.65668483°$$

$$\theta = 44.7°$$

51.

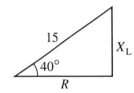

$$\sin 40° = \frac{\text{opp}}{\text{hyp}}$$

$$\sin 40° = \frac{X_L}{15}$$

$$(15)\sin 40° = X_L$$

$$X_L = 15(0.6427876097)$$

$$X_L = 9.641814145$$

$$X_L = 9.64 \ \Omega$$

$$\cos 40° = \frac{\text{adj}}{\text{hyp}}$$

$$\cos 40° = \frac{R}{15}$$

$$R = 15 \cos 40°$$

$$R = 15(0.7660444431)$$

$$R = 11.49066665$$

$$R = 11.49 \ \Omega$$

Chapter 20 Practice Test

1.

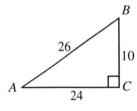

$$\sin A = \frac{\text{opp}}{\text{hyp}} = \frac{10}{26} = \frac{5}{13}$$

3.

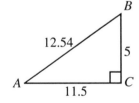

$$\cos A = \frac{\text{adj}}{\text{hyp}} = \frac{11.5}{12.54} = 0.9171$$

5. $\sin 53° = 0.7986$

7. $\sin \theta = 0.2756$
$\theta = \sin^{-1}(0.2756)$
$\theta = 16.0°$

9. $\sin \theta = 0.7660$
$\theta = \sin^{-1}(0.7660)$
$\theta = 0.8726$

11.

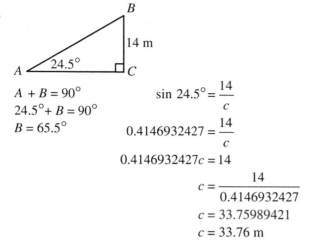

$A + B = 90°$
$24.5° + B = 90°$
$B = 65.5°$

$\sin 24.5° = \dfrac{14}{c}$

$0.4146932427 = \dfrac{14}{c}$

$0.4146932427c = 14$

$c = \dfrac{14}{0.4146932427}$

$c = 33.75989421$

$c = 33.76$ m

$a^2 + b^2 = c^2$
$14^2 + b^2 = 33.76^2$
$196 + b^2 = 1139.7376$
$b^2 = 943.7376$
$b = \sqrt{943.7376}$
$b = 30.7203125$
$b = 30.72$ m

13.

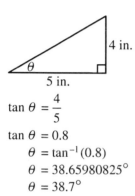

$\tan \theta = \dfrac{4}{5}$

$\tan \theta = 0.8$

$\theta = \tan^{-1}(0.8)$

$\theta = 38.65980825°$

$\theta = 38.7°$

15.

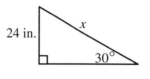

$\sin 30° = \dfrac{24}{x}$

$0.5 = \dfrac{24}{x}$

$0.5x = 24$

$x = \dfrac{24}{0.5}$

$x = 48$ in.

17.

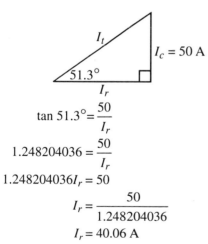

$\tan 51.3° = \dfrac{50}{I_r}$

$1.248204036 = \dfrac{50}{I_r}$

$1.248204036 I_r = 50$

$I_r = \dfrac{50}{1.248204036}$

$I_r = 40.06$ A

19.

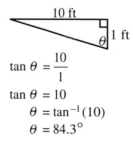

$\tan \theta = \dfrac{10}{1}$

$\tan \theta = 10$

$\theta = \tan^{-1}(10)$

$\theta = 84.3°$

chapter 21 Trigonometry with Any Angle

Chapter Review Exercises

1. (4, 6)

magnitude	direction
$p^2 = x^2 + y^2$	$\tan \theta = \dfrac{y}{x}$
$p^2 = 4^2 + 6^2$	
$p^2 = 16 + 36$	$\tan \theta = \dfrac{6}{4}$
$p^2 = 52$	
$p = \sqrt{52}$	$\tan \theta = 1.5$
$p = 7.211102551$	$\theta = \tan^{-1}(1.5)$
$p = 7.211$	$\theta = 56.30993247°$
	$\theta = 56.3°$

3. $(1 + 4j) + (6 + 8j) = 7 + 12j$

magnitude	direction
$p^2 = x^2 + y^2$	$\tan \theta = \dfrac{y}{x}$
$p^2 = 7^2 + 12^2$	
$p^2 = 49 + 144$	$\tan \theta = \dfrac{12}{7}$
$p^2 = 193$	
$p = \sqrt{193}$	$\tan \theta = 1.714285714$
$p = 13.89244399$	$\theta = \tan^{-1}(1.714285714)$
$p = 13.89$	$\theta = 59.74356284°$
	$\theta = 59.7°$

5. $3.5(2.4) = 8.4$ magnitude

$\theta = 45°$ direction (same as original vector)

The resultant vector has a magnitude of 8.4 and a direction of 45°

7. (3, 5.2) $\theta = \tan^{-1} \dfrac{y}{x}$

$r = \sqrt{x^2 + y^2}$

$r = \sqrt{3^2 + 5.2^2}$ $\theta = \tan^{-1} \dfrac{5.2}{3}$

$r = \sqrt{9 + 27.04}$ $\theta = \tan^{-1} 1.733333333$

$r = \sqrt{36.04}$

$r = 6.0$ $\boxed{\text{TAN}^{-1}}$ 1.733333333 $\boxed{\text{ENTER}}$ =>

$\theta = 1.047518004$

Polar coordinates are (6.0, 1.0)

9. x-coordinate: $\cos \theta = \dfrac{x}{r}$ y-coordinate: $\sin \theta = \dfrac{y}{r}$ Rectangular coordinates: (4.2, 10.2)

$\cos \dfrac{3\pi}{8} = \dfrac{x}{11}$ $\sin \dfrac{3\pi}{8} = \dfrac{y}{11}$

$x = 11 \cos \dfrac{3\pi}{8}$ $y = 11 \sin \dfrac{3\pi}{8}$

$x = 11(0.3826834324)$ $y = 11(0.9238795325)$

$x = 4.209517756$ $y = 10.16267486$

11.   related angle of 3.04 rad

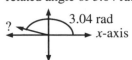

3.1415926540 rad (π)
$-$ 3.0400000000 rad
_____
0.1015926536 rad
= 0.1016 rad

13.   related angle of 221°

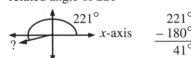

221°
$-$ 180°
_____
41°

15.   related angle of 5.4 rad

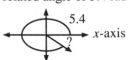

6.283185307 rad (2π)
$-$ 5.400000000 rad
_____
0.883185307 rad
= 0.8832 rad

17.   related angle of 212°15′10″

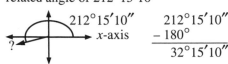

212°15′10″
$-$ 180°
_____
32°15′10″

19. $\sin 2.1 = 0.8632093666 = 0.8632$

21. $\sin 340° = -0.3420201433 = -0.3420$

23. $\cos 290° = 0.3420201433 = 0.3420$

25. $\tan \dfrac{5\pi}{4} = 1$

27. $(-2, 2)$

magnitude
$p^2 = x^2 + y^2$
$p^2 = (-2)^2 + (2)^2$
$p^2 = 4 + 4$
$p^2 = 8$
$p = \sqrt{8}$
$p = 2.828427125$
$p = 2.83$

direction
$\tan \theta = \dfrac{y}{x}$

$\tan \theta = \dfrac{2}{-2}$

$\tan \theta = -1$
$\theta = \tan^{-1}(-1)$
$\theta = -0.7853981634$ rad   Reference angle must be in quadrant II.
$\theta = \pi - |\theta_2|$
π rad $- 0.7853981634 = 2.35619449$ rad or 2.36 rad

29. $T = \dfrac{1}{f}$

$T = \dfrac{1}{15,000 \text{ Hz}}$

$T = 0.000067$ s

31. $y = 5 \cos 4x$
In the equation, $y = A \cos Bx$, A represents the amplitude. $A = 5$.

Period
$P = \dfrac{360°}{B}$
$P = \dfrac{360°}{4}$
$P = 90°$

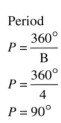

33. $v = V \sin x$
$v = 50 \sin 120°$
$v = 43.3$

35. $B = 120°$, $C = 20°$, $a = 8$ m

$$A + B + C = 180°$$
$$A + 120° + 20° = 180°$$
$$A + 140° = 180°$$
$$A = 180° - 140°$$
$$A = 40°$$

$$\frac{a}{\sin A} = \frac{b}{\sin B}$$
$$\frac{8}{\sin 40°} = \frac{b}{\sin 120°}$$
$$b = \frac{8 \sin 120°}{\sin 40°}$$
$$b = 10.77837084$$
$$b = 10.8 \text{ m}$$

$$\frac{a}{\sin A} = \frac{c}{\sin C}$$
$$\frac{8}{\sin 40°} = \frac{c}{\sin 20°}$$
$$c = \frac{8 \sin 20°}{\sin 40°}$$
$$c = 4.25671109$$
$$c = 4.3 \text{ m}$$

37. $a = 5$, $c = 7$, $C = 45°$

$$\frac{a}{\sin A} = \frac{c}{\sin C}$$
$$\frac{5}{\sin A} = \frac{7}{\sin 45°}$$
$$\sin A = \frac{5 \sin 45°}{7}$$
$$\sin A = 0.5050762723$$
$$A = \sin^{-1}(0.5050762723)$$
$$A = 30.33641562°$$
$$A = 30.3°$$

$$A + B + C = 180°$$
$$30.3° + B + 45° = 180°$$
$$B + 75.3° = 180°$$
$$B = 180° - 75.3°$$
$$B = 104.7°$$

$$\frac{b}{\sin B} = \frac{c}{\sin C}$$
$$\frac{b}{\sin 104.7°} = \frac{7}{\sin 45°}$$
$$b = \frac{7 \sin 104.7°}{\sin 45°}$$
$$b = 9.575462221$$
$$b = 9.6 \text{ dkm}$$

39. Since we are given a, b, and B, and because $b < a$, there are 2 solutions.
$a = 9.2$, $b = 6.8$, $B = 28°$

$$\frac{a}{\sin A} = \frac{b}{\sin B}$$
$$\frac{9.2}{\sin A} = \frac{6.8}{\sin 28°}$$
$$\sin A = \frac{9.2 \sin 28°}{6.8}$$
$$\sin A = 0.6351674085$$
$$A = \sin^{-1}(0.6351674085)$$
$$A = 39.43240195°$$
$$A = 39.4°$$
or
$$A = 180° - 39.4° = 140.6°$$

$$A + B + C = 180°$$
$$39.4° + 28° + C = 180°$$
$$C + 67.4° = 180°$$
$$C = 180° - 67.4°$$
$$C = 112.6°$$

$$\frac{b}{\sin B} = \frac{c}{\sin C}$$
$$\frac{6.8}{\sin 28°} = \frac{c}{\sin 112.6°}$$
$$c = \frac{6.8 \sin 112.6°}{\sin 28°}$$
$$c = 13.37211873$$
$$c = 13.4 \text{ cm}$$

Solution 1:
$$\boxed{\begin{array}{l} A = 39.4° \\ C = 112.6° \\ c = 13.4 \text{ cm} \end{array}}$$

$$A + B + C = 180°$$
$$140.6° + 28° + C = 180°$$
$$C + 168.6° = 180°$$
$$C = 180° - 168.6°$$
$$C = 11.4°$$

$$\frac{b}{\sin B} = \frac{c}{\sin C}$$
$$\frac{6.8}{\sin 28°} = \frac{c}{\sin 11.4°}$$
$$c = \frac{6.8 \sin 11.4°}{\sin 28°}$$
$$c = 2.862942127$$
$$c = 2.9 \text{ cm}$$

Solution 2:
$$\boxed{\begin{array}{l} A = 140.6° \\ C = 11.4° \\ c = 2.9 \text{ cm} \end{array}}$$

41.   Find *RS*. There is only one case since 29 ft ≥ 20 ft.

T

29 ft   20 ft

R 38° S

let $t = RS$

$$\frac{t}{\sin T} = \frac{s}{\sin S}$$

$T + S + R = 180°$

$T + 38° + 25.1° = 180°$

$T + 63.1° = 180°$

$T = 180° - 63.1°$

$T = 116.9°$

$$\frac{t}{\sin 116.9°} = \frac{29}{\sin 38°}$$

$$t = \frac{29 \sin 116.9°}{\sin 38°}$$

$t = 42.00705972$ ft

$RS = t = 42.0$ ft

43. A

16   115°

C 115° B

18

$$\frac{a}{\sin A} = \frac{c}{\sin C}$$

$$\frac{18}{\sin A} = \frac{28.69543725}{\sin 115°}$$

$$\sin A = \frac{18 \sin 115°}{28.69543725}$$

$\sin A = 0.5685064153$

$A = \sin^{-1}(0.5685064153)$

$A = 34.64613909$

$A = 34.6°$

$c^2 = a^2 + b^2 - 2ab \cos C$

$c^2 = 18^2 + 16^2 - 2(18)(16)\cos 115°$

$c^2 = 324 + 256 - 576(-0.4226182617)$

$c^2 = 324 + 256 + 243.4281188$

$c^2 = 823.4281188$

$c = \sqrt{823.4281188}$

$c = 28.69543725$

$c = 28.70$

$A + B + C = 180°$

$34.6° + B + 115° = 180°$

$B + 149.6° = 180°$

$B = 180° - 149.6°$

$B = 30.4°$

45. A

27   34°   27

B C

$a^2 = b^2 + c^2 - 2bc \cos A$

$a^2 = 27^2 + 27^2 - 2(27)(27)\cos 34°$

$a^2 = 729 + 729 - 1{,}458(0.8290375726)$

$a^2 = 729 + 729 - 1{,}208.7360781$

$a^2 = 249.2632192$

$a = \sqrt{249.2632192}$

$a = 15.78807206$

$a = 15.79$

$B = C$ since the triangle is isosceles

$A + B + C = 180°$

$34° + 2B = 180°$

$2B = 146°$

$B = 73°$

$C = 73°$

47. A

8.2

B 60° C

8.2

$A = C$ since the angles opposite the equal sides are equal.

$A + B + C = 180°$

$A + 60° + C = 180°$

$A + C = 120°$

$A + A = 120°$

$2A = 120°$

$A = 60° = C$

therefore, the triangle is an equilateral triangle

so $b = 8.2$

49.

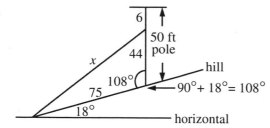

$a^2 = b^2 + c^2 - 2bc \cos A$

$x^2 = 44^2 + 75^2 - 2(44)(75)\cos 108°$

$x^2 = 1,936 + 5,625 - 6,600(-0.3090169944)$

$x^2 = 1,936 + 5,625 + 2,039.512163$

$x^2 = 9,600.512163$

$x = \sqrt{9,600.512163}$

$x = 97.98 \text{ ft}$

51. Find x and y.

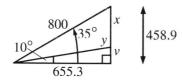

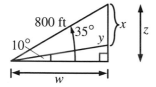

small right triangle

$\tan 10° = \dfrac{v}{655.3216354}$

$v = 655.3216354 \tan 10°$

$v = 115.5508854$

$v = 115.6$

$x = z - v$

$x = 458.9 - 115.6$

$x = 343.3 \text{ ft vertical feet to be removed}$

large right triangle

$\sin 35° = \dfrac{z}{800}$

$z = 800 \sin 35°$

$z = 458.8611491$

$z = 458.9 \text{ ft}$

$w^2 + z^2 = 800^2$

$w^2 + 458.8611491^2 = 800^2$

$w^2 + 210,553.5541 = 640,000$

$w^2 = 429,446.4459$

$w = \sqrt{429,446.4459}$

$w = 655.3216354$

$w = 655.3 \text{ ft}$

$w^2 + v^2 = y^2$

$655.3216354^2 + 115.5508854^2 = y^2$

$429,446.4458 + 13,352.00712 = y^2$

$442,798.4529 = y^2$

$\sqrt{442,798.4529} = y$

$y = 665.4310279$

$y = 665.4 \text{ ft}$

distance from bottom to top of new road bed

Chapter 21 Practice Test

1. $\sin 125° = 0.8191520443 = 0.8192$

3. $\cos 160° = -0.9396926208 = -0.9397$

5. $(-8, 8)$

$\tan \theta = \dfrac{y}{x}$ $\theta = \tan^{-1}(-1)$

$\theta = -45°$

$\tan \theta = \dfrac{8}{-8}$ related angle is $45°$

quadrant II

$\tan \theta = -1$ $180° - 45° = 135°$

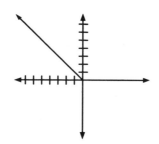

7.

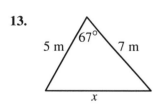

14 ft   16.5 ft
θ   35.5°

$35.5° + 43.2° + x = 180°$
$x + 78.7° = 180°$
$x = 180° - 78.7°$
$x = 101.3°$

$$\frac{14}{\sin 35.5°} = \frac{16.5}{\sin \theta}$$
$$\sin \theta = \frac{16.5 \sin 35.5°}{14}$$
$$\sin \theta = 0.6843999121$$
$$\theta = \sin^{-1}(0.6843999121)$$
$$\theta = 43.18843189°$$
$$\theta = 43.2°$$

9.

12 ft   12 ft
68°
x

isosceles triangle with equal angles opposite equal sides, both base angles = 68°
$68° + 68° + \theta = 180°$
$\theta + 136° = 180°$
$\theta = 180° - 136°$
$\theta = 44°$
$$\frac{x}{\sin 44°} = \frac{12}{\sin 68°}$$
$$x = \frac{12 \sin 44°}{\sin 68°}$$
$$x = 8.990558242 \text{ ft}$$
$$x = 8.991 \text{ ft}$$

11.

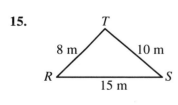

11 ft   21 ft
x
12 ft

$21^2 = 11^2 + 12^2 - 2(11)(12)\cos x$
$441 = 121 + 144 - 264\cos x$
$176 = -264\cos x$
$-0.6666666667 = \cos x$
$x = \cos^{-1}(0.6666666667)$
$x = 131.8103149°$
$x = 131.8°$

13.

5 m   67°   7 m
x

$x^2 = 5^2 + 7^2 - 2(5)(7)\cos 67°$
$x^2 = 25 + 49 - 70(0.3907311285)$
$x^2 = 74 - 27.351017899$
$x^2 = 46.64882101$
$x = 6.829994217 \text{ m}$
$x = 6.830 \text{ m}$

15.

T
8 m   10 m
R   15 m   S

$r^2 = s^2 + t^2 - 2st\cos R$
$$\cos R = \frac{s^2 + t^2 - r^2}{2st}$$
$$\cos R = \frac{8^2 + 15^2 - 10^2}{2(8)(15)}$$
$$\cos R = \frac{8^2 + 15^2 - 10^2}{240}$$
$$\cos R = \frac{64 + 225 - 100}{240}$$

$$\cos R = \frac{189}{240} = 0.7875$$
$R = \cos^{-1} 0.7875$
$R = 38.04750745°$
$R = 38.0°$

17. $a^2 = 76^2 + 110^2 - 2(76)(110)\cos 107°$
$a^2 = 5{,}776 + 12{,}100 - 16{,}720(-0.2923717047)$
$a^2 = 5{,}776 + 12{,}100 + 4{,}888.454903$
$a^2 = 22{,}764.4549$
$a = 150.8789412$ ft
$a = 150.9$ ft

19. $a^2 = b^2 + c^2 - 2ac\cos A$
$a^2 = 12^2 + 15^2 - 2(12)(15)\cos 48°$
$a^2 = 144 + 225 - 360(0.6691306064)$
$a^2 = 128.1129817$
$a = \sqrt{128.1129817}$
$a = 11.31870053$ ft
$a = 11.32$ ft

21.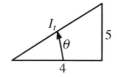
$I_t^2 = 4^2 + 5^2$
$I_t^2 = 16 + 25$
$I_t^2 = 41$
$I_t = 6.4$ milliamps

23. $\cos 60° = \dfrac{15}{Z}$
$Z\cos 60° = 15$
$Z = \dfrac{15}{\cos 60°}$
$Z = 30$ ohms

Chapters 20–21 Cumulative Practice Test

1. (9, 12)
$r = \sqrt{x^2 + y^2}$
$r = \sqrt{9^2 + 12^2}$
$r = \sqrt{81 + 144}$
$r = \sqrt{225}$
$r = 15$
$\theta = \tan^{-1}\dfrac{y}{x}$
$\theta = \tan^{-1}\dfrac{12}{9}$
$\theta = \tan^{-1} 1.333333333$
$\theta = 53.1°$
Polar coordinates: (15, 53.1°)

3. (2.7, 3.5)
$r = \sqrt{x^2 + y^2}$
$r = \sqrt{2.7^2 + 3.5^2}$
$r = \sqrt{7.29 + 12.25}$
$r = \sqrt{19.54}$
$r = 4.4$
$\theta = \tan^{-1}\dfrac{y}{x}$
$\theta = \tan^{-1}\dfrac{3.5}{2.7}$
$\theta = \tan^{-1} 1.296296296$
$\theta = 52.4°$
Polar coordinates: (4.4, 52.4°)

5. $(7.4, 60°)$

x-coordinate: $x = r \cos \theta$

$$x = 7.4 \cos 60°$$
$$x = 3.7$$

y-coordinate: $y = r \sin \theta$

$$y = 7.4 \sin 60°$$
$$y = 6.4$$

Rectangular coordinates: $(3.7, 6.4)$

7. $\left(4.8, \dfrac{\pi}{8}\right)$

x-coordinate: $\cos \theta = \dfrac{x}{r}$

$$\cos \frac{\pi}{8} = \frac{x}{4.8}$$
$$4.8 \cos \frac{\pi}{8} = x$$
$$x = 4.8(0.9238795325)$$
$$x = 4.434621756$$

y-coordinate: $\sin \theta = \dfrac{y}{r}$

$$\sin \frac{\pi}{8} = \frac{y}{4.8}$$
$$4.8 \sin \frac{\pi}{8} = y$$
$$4.8(0.3826834304) = y$$
$$y = 1.836880475$$

$(4.4, 1.8)$

9. $\cos A = \dfrac{\text{adj}}{\text{hyp}}$

$$\cos 60° = \frac{23.5 \text{ cm}}{\text{hyp}}$$
$$\text{hyp} = \frac{23.5 \text{ cm}}{\cos 60°}$$
$$\text{hyp} = \frac{23.5 \text{ cm}}{0.5}$$
$$AB = 47 \text{ cm}$$

11. $\sin B = \dfrac{\text{opp}}{\text{hyp}}$

$$\sin B = \frac{12.5 \text{ cm}}{15.8 \text{ cm}}$$
$$\sin B = 0.7911392405$$
$$B = \sin^{-1}(0.7911392405)$$
$$B = 52.292100309°$$
$$\angle B = 52.3°$$

13.

Magnitude	Direction
$c = \sqrt{a^2 + b^2}$	$\tan \theta = \dfrac{\text{opp}}{\text{adj}}$
$c = \sqrt{5^2 + 2^2}$	
$c = \sqrt{25 + 4}$	$\tan \theta = \dfrac{2}{5}$
$c = \sqrt{29}$	$\tan \theta = 0.4$
$c = 5.385164807$	$\theta = \tan^{-1}(0.4)$
magnitude = 5.385	$\theta = 21.80140949°$
	$\theta = 21.8°$
	direction = $21.8°$

15. $y = \cos(x + 90°)$

$$y = A \cos(Bx + C)$$

Amplitude $(A) = 1$

$$\text{Period} = \frac{360°}{B} = \frac{360°}{1} = 360°$$

$$\text{Phase shift} = \frac{C}{B} = \frac{90°}{1} = 90°$$

Sign is *positive*; so phase shift is $90°$ *left*.

17. $y = 3\sin(2x - 240°)$

$y = A\cos(Bx + C)$

Amplitude $(A) = 3$

Period $= \dfrac{360°}{B} = \dfrac{360°}{2} = 180°$

Phase shift $= \dfrac{C}{B} = \dfrac{-240°}{2} = -120°$

Sign is *negative*; so phase shift is $120°$ *right*.

19. $c^2 = a^2 + b^2 - 2ab\cos C$

$c^2 = 36^2 + 34^2 - 2(36)(34)\cos 38°$

$c^2 = 1{,}296 + 1{,}156 - 2{,}448(0.7880107536)$

$c^2 = 1{,}296 + 1{,}156 - 1{,}929.050325$

$c^2 = 522.9496752$

$c = \sqrt{522.9496752}$

$c = 22.86809295$

$AB = 23$ cm (rounded)

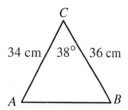